THE BURNS AND LAIRD FAMILY INTERESTS
in the
FORMATION OF COAST LINES

Nick Robins

Malcolm McRonald

CONTENTS

PREFACE	3
1. SAIL AND STEAM	4
2. WOODEN PADDLE STEAMERS BETWEEN GLASGOW, LONDONDERRY AND SLIGO	10
3. ENTER MR ALEXANDER A LAIRD JUNIOR	15
4. EXIT MR CAMERON	21
5. ALEXANDER A LAIRD IN CHARGE	26
6. REBRANDING AND 'THE LAIRD'S LINE'	33
7. GROWING TRADE UP UNTIL 1904	39
8. LAIRD LINE LIMITED	49
9. J & G BURNS; JAMES MARTIN AND J & G BURNS; GEORGE BURNS – SHIP OWNERS	58
10. G & J BURNS' DIVERSE TRADE INTERESTS	64
11. CONSOLIDATION	70
12. SCREW STEAMERS TO BELFAST	77
13. DAYLIGHT AT LAST	83
14. G & J BURNS LIMITED	93
15. WAR, PEACE, TAKEOVER AND MERGER	100
REFERENCES	107
FLEET LIST	108
INDEX	134

PREFACE

Although the history of Burns & Laird Lines Limited is reasonably well known, no detailed account of its predecessors up until their acquisition by Coast Lines has previously been written. Both the Burns and the Laird interests go back to the very early days of the steamship. J & G Burns were keenly involved with the Glasgow to Belfast and the Glasgow to Liverpool services. The Burns brothers were also involved in Clyde and West Highland services until these were split away into a new and separate company, later to become David MacBrayne Limited. The Laird interests focussed more on routes from Glasgow to the north and north west of Ireland, making Londonderry the key destination. While the Burns family also became founders and leaders of the emerging Cunard Steamship Company, Alexander Laird stayed in the Irish trade to develop additional routes to Dublin and between Morecambe and Ireland.

The story of the development of both the Burns and the Laird companies during the reign of Victoria is colourful and exciting. The business of the companies peaked before the Great War when it began to suffer from a downturn in trade caused by political unrest in Ireland. Loss of many of the passenger vessels in the Great War, coupled with post-war recession, allowed a fire-sale of both companies to Coast Lines Limited. This is the story of the successes and failures, the triumphs and traumas, of both the Burns and the Laird interests on the Irish Sea before the merger of the companies in 1922.

A variety of information sources has been consulted in preparing this history. These include the National Archives and local Record Offices, the Merseyside Maritime Museum, shipping registers held at Glasgow and Liverpool, the Mercantile Navy List, Lloyd's Register and the British Newspaper Archive. The authors are grateful for all the help given in accessing data and to all those individuals who have pointed us towards other information. Historic photographic material is either from Nick Robins' collection or from the late Mike Walker's collection of postcards and photographs. This book is dedicated to the memory of Mike Walker. Mike was a gentleman with a story to tell about almost every cross-channel ship that ever sailed the Irish Sea. His knowledge of railway shipping around the whole of the British Isles was exceptional; Mike, we are grateful to you for having shared some of that understanding with us.

As always we are grateful to Gil Mayes for proof reading and to Bernard McCall and his team at Portishead for creating the finished product to his customary high standard.

Nick Robins, Crowmarsh, and Malcolm McRonald, Heswall   August 2018

First published in 2018 by

Coastal Shipping Publications
400 Nore Road, Portishead, BS20 8EZ, England
Tel: +44(0)1275 846178
Email bernard@coastalshipping.co.uk
Website: www.coastalshipping.co.uk

ISBN: 978-1-902953-90-8

Copyright Nick Robins and Malcolm McRonald

The rights of Nick Robins and Malcolm McRonald to be identified as authors of this work have been asserted by them in accordance with the Copyright, Design and Patent Act 1988. All rights reserved.
No part of this publication may be reproduced, stored in a retrieval system or transmitted in any form or by any means (electronic, digital, mechanical, photocopying, recording or otherwise) without prior permission of the publisher.
All distribution enquiries should be addressed to the publisher.

Printed by Short Run Press
25 Bittern Road, Exeter, EX2 7LW
Tel: +44(0)1392 211909 Fax: +44(0)1392 444134
Email: books@shortrunpress.co.uk
Website: www.shortrunpress.co.uk

Front cover: 'Glasgow Boat, Portrush harbour' the *Azalea* (1878) [Mike Walker collection]
Back cover: Cover of the 1902 timetable for the Ardrossan and Belfast daylight service by the paddle steamer *Adder*

CHAPTER 1

SAIL AND STEAM

The number of merchant steamships in the Home Trade took a long time to overtake sail once steam was introduced in 1814 to estuarine waters. *The Mercantile Navy List* (1862) tells us that forty years later, in 1854, there were 8,538 sailing ships occupied full time in the Home Trade and just 240 steam vessels. However, the average burthen of the sailing ships was only 81 tons whereas that for the steamships was a much larger 225 tons. Such economies of scale meant that the steamship must one day overtake the sailing ship, not just in the Home Trade but also in the Foreign Trade.

In the early days of steam the reliability of the double-acting single cylinder steam engine mounted aboard a small wooden hull and driving fixed wooden paddle floats was poor. However, the concept of steam was soon accepted by the travelling public, especially by passengers on coastal and cross-channel services. It was not long before the steamship made inroads into the business that until then had been maintained solely by sailing smacks and schooners. The reason was simple – the travellers knew they would leave Greenock one day and, for the most part, they would land at Belfast the next. That the service was erratic and irregular was, at that stage in the evolution of the steamship, of little concern. As the hulls became bigger, so the engines increased in weight and became more powerful and more reliable; only then did the passenger begin to appreciate a sailing timed for 7pm with arrival scheduled for the next afternoon.

The lead in paddle steamer engineering in the UK came firmly from the Clyde. It was not surprising, therefore, that the first significant uptake in cross-channel steamers should be on the route between the Clyde and Belfast. The first regular cross-channel steamer in service was the **Rob Roy** as reported in the *Glasgow Herald*, Friday 26 June 1818:

> We are happy to announce the safe arrival of the **Rob Roy** steam packet from Dublin. On Friday morning [19 June] at 4 o'clock she sailed from the Broomielaw, touched at Port Glasgow, Greenock and Gourock; but owing to the boisterous state of the weather, it was deemed advisable to put into Lamlash Bay, where she lay for eight hours. She proceeded to Carrickfergus and arrived in Dublin on Sunday morning at 3 o'clock. At the earnest request of the passengers, Mr Napier the proprietor, kindly consented to remain in Dublin two days, Sunday and Monday and on Tuesday morning [23 June] at 5 o'clock, she sailed for Greenock where she arrived at 7 o'clock on Wednesday morning, having performed the passage in an unprecedented, short time of 26 hours.

The little steamer had earlier completed a round trip from Glasgow to Belfast calling at Campbeltown on 13 June, and returning to Greenock two days later. The **Rob Roy** was able to maintain 7 knots and was a match for the sailing smacks that served the route year round. After a few more trial runs to Dublin, with calls in Belfast Lough, she settled down to a twice weekly roster from Greenock to Belfast via Campbeltown returning directly to Greenock. The service was suspended in the autumn when the **Rob Roy** received a complete machinery overhaul and upgrade of her accommodation to include separate dormitory style berths for men and women. The **Rob Roy** remained in seasonal employment between the Clyde and Belfast for the next three years, retiring to lay up at Greenock during the winter when the smacks regained supremacy. Early in the 1821 season she was put to use on the English Channel, her role in demonstrating the viability of the paddle steamer on the Greenock to Belfast route being complete. *Memoirs and Portraits of One Hundred Glasgow Men* includes the accolade:

> The **Rob Roy** (90 tons and 30 H.P.) was built by William Denny of Dumbarton, and engined by David Napier himself. Proving too small for the Belfast trade, she was transferred to the Dover and Calais route. She had been the first steamer between Great Britain and Ireland: she was now the first steamer between Great Britain and the Continent; and she ended by being the first steamer owned by the French Government. They were so pleased with the way she did her work that they bought her, and, with the help of a priest and holy water, changed **Rob Roy** into **Henry Quatre.**

Her successor was the **Eclipse**, advertised to sail from the Broomielaw on Wednesdays and Saturdays, returning from Belfast on Mondays and Thursdays. Her 'commander' was Captain James Dalzell, and it was under him that this little steamer achieved the passage to Belfast in just under twelve hours. The ship was popular and in October 1821 she carried an exceptional 250 passengers on one voyage.

The early steamers were now set to colonise the Irish Sea with services between Liverpool and Dublin, as well as coasting services such as that between Greenock and Liverpool. The opportunities were tremendous, the risks were high and the expenses of maintaining a steamship for a regular scheduled service were massive. It was to be a long while before all the sailing ships plying between the Clyde and Ireland, and on the coasting services south to Liverpool, were completely ousted from profitable employment. Many bulk cargoes, especially coal landed at small harbours and onto beaches, continued to rely on sail as did the numerous non-perishable and low value cargoes. The fast so-called 'rantapikes' sailing between Glasgow and Liverpool continued to take the steamers head on and offered lower freight rates but less predictable, and sometimes faster, passage times than the steamers. The last of the sailing ships on this route was commissioned by Robert Gilchrist as late as the 1850s.

One notable pair of brothers willing to accept the risks of early steam navigation were the Burns brothers, James and George. They were passionate about developing the steamship as a radical improvement over sail. The Burns brothers were well prepared for the challenge of ship owning and they belonged to a family of considerable influence in various spheres of Glasgow life. Their father, the Reverend Dr John Burns, was minister of the Barony Parish in Glasgow. As it happened, their sister Elizabeth was to become the mother of the future West Highland entrepreneur, David MacBrayne. Dr Burns had nine children, including two notable surgeons, John and Allan; the former died in the tragic loss of the G & J Burns steamer **Orion** (Chapter 10) while the latter died at an early age due to an accident whilst he was carrying out an operation on a patient.

Memoirs and Portraits of One Hundred Glasgow Men again:

> It was in 1818 when the two brothers [James and George] became partners. J & G Burns, their original firm, were produce merchants. In pushing their business they visited Belfast and every Irish port, as well as Liverpool and London, and everywhere the young Scotch firm made friends. With the help of these friends the firm soon promised to take a leading place in the produce line. But their energies were turned into another channel, and the main outcome of the produce business proved to be the help, which the friends had won them, given also to the Burns's in their new line of shipping. Their connection with shipping dates from 1824. They began with the Liverpool trade, and have been unremitting in their attentions to their first love.

> …The Glasgow and Liverpool trade was divided among three companies, owning in all eighteen smacks. They sometimes made the passage to Liverpool (of course from Greenock) in 24 hours; the average passage was three to four days.

> Six smacks were owned by a Glasgow joint-stock company; the agent here was James Martin, who had been up till 1814 in the West India trade; his brother Thomas Martin was agent in Liverpool. Six smacks were owned by a private Liverpool firm Matthie & Theakstone, both able men; the senior partner was Hugh Matthie, an active, shrewd old Scotchman; the owners were their own agents in Liverpool; the agents here [Glasgow] were two brothers, Port-Dundas men, John and Alexander Kidd, with a young Port-Glasgow man, David Hutcheson, as chief clerk. The remaining six smacks belonged to a private Glasgow company, managed here by David Chapman, afterwards of Thomson & McConnell, and in Liverpool by William Swan Dixon, one of the Dumbarton Dixons.

> In 1824 the second of the two Kidds died, and J & G Burns applied for the agency. They had strong competitors in Fleming & Hope, another produce firm, older and better known, and backed by a formidable round-robin. The choice lay with Hugh Matthie. The Burns's had come repeatedly in contact with him about freight for their produce, and had left on him their usual good impression; and personal fitness weighed more with Hugh than round-robins: but he would come down and judge for himself. He came down, and he chose the Burns's. He gave them full power, but suggested that they might retain David Hutcheson, whose usefulness he had noticed after the death of the Kidds. The Burns's engaged Hutcheson, and took new premises for the new business. James Burns stuck to the produce, and as long as he lived this branch was kept up, though it had long been little but the victualling department of the Burns fleet. George Burns at once threw himself into the shipping.

> ...Very soon after the Burns's became agents, Theakstone wished to retire, and the Burns's bought his share, thus becoming equal partners with Matthie, and for the first time shipowners…

The Burns' interest as ship owners was a half share in the six sailing smacks, the ownership being registered in the name of George Burns. It is notable that there were so many recognisable names in the shipping industry of that era, reflecting the tight knit community of Glasgow-based ship owners.

It was not long before the Burns brothers grasped the issue of adopting steamships in place of the smacks, *Memoirs and Portraits of One Hundred Glasgow Men* again:

> As soon as the Burns's saw that the future was with steam, they urged its adoption on Hugh Matthie, their partner, and on James Martin for his company. As they put it, the clearing away of their twelve smacks would make good room for steamers. Matthie was too old and too rich to care for the new venture, but he kindly yielded to the Burns influence, only recommending that somewhere half-way there should be a reserve depot for coals - for Hugh's own faith in steamers was weak. Martin was easily won over, but only after great trouble and delay got his owners to agree. The two agencies were then amalgamated under the lumbering title of James Martin & J. & G. Burns, the half-way coals were duly stored at Port-Nessock to crumble peacefully away, the smacks vanished, and on Friday, the 13th March, 1829, the first steamer of the new Glasgow company passed the Cloch. She was appropriately named the **Glasgow**, and was followed next month by the **Ailsa Craig**, a crack vessel of her day, and next year by the **Liverpool**, of which old Hugh Main was captain.

These new brooms swept so clean that they left little for the old ones. For some time the Manchester men stood their punishment, but when the captain of the *James Watt* reported to her owners that, on his way to Glasgow, he had met the *Ailsa Craig*, which had left Liverpool about the same time with him, steaming merrily back again, they threw up the sponge, and offered to hand over the whole concern to the Glasgow company: meantime, and preliminary to this, they placed their steamers and plant on commission under that company's agency. Difficulties among themselves stopped the carrying out of the offer, but the Manchester company withdrew their vessels, and Burgh Reeve Burt and the leading partners promised the Glasgow company all their influence against any opposition.

The 'Manchester company', correctly titled the Mersey and Clyde Steam Navigation Company, had operated the **Henry Bell, James Watt** and **William Huskisson**. The owners were a powerful group of Manchester men including Mr Burt, the Burgh Reeve. The Greenock agent for this company was Alexander Allan Laird, whose son, also named Alexander Allan Laird (junior), would one day generate the Laird Line. This company was destined to merge with the Burns' enterprise in 1922, albeit under Coast Lines' ownership, to become Burns & Laird Lines Limited.

Alexander Laird (senior) was an established agent at Greenock whose main task was to ensure that cargoes were delivered to and collected from the port and that freight dues were fully paid. For example, in June 1813 Alexander Laird is recorded in the Decisions of the Court of Session as port agent for ship owner George Malcolm as well as the timber merchant Allan Scott & Company. The importer apparently went bankrupt when half of the timber cargo had been delivered from the ship and a court case ensued over payment of dues to George Malcolm, following Laird's assurance that the remainder of the cargo could safely be delivered.

Alexander Laird (senior) first came to prominence as a shipbroker in 1822 when he was appointed agent for the Saint George Steam Packet Company at that port. That company's first steamer, **Saint George**, ran a summer service between Liverpool and Greenock via Douglas in competition with James Little's service, which had started in 1819 with the **Majestic** and **City of Glasgow**. During this first year, the **Saint George** was transferred to Liverpool to Dublin duties and Laird lost his agency for the company. In 1823 the Laird agency used this initial experience to win the agency for the newly created Mersey and Clyde Steam Navigation Company which then advertised a twice weekly service to Liverpool with the new steamers **Henry Bell** and **James Watt** sailing from Greenock at 6 pm on Tuesdays and Saturdays. 'Excellent Tables served on board at low rates, and the Cabins are well supplied with Newspapers and choice libraries.' The single cabin passage fare was 21/-, although sailings were suspended from the end of November until the beginning of April.

Laird's son joined the agency business, now trading as Alex Laird & Company, in 1824 as an apprentice. Alexander Allan Laird (Junior) was born in the Gorbals, Glasgow on 22 March 1811 to Alexander Allan Laird and Catherine neé McCauley. The focus of business for Laird (senior) was fast developing at Greenock. The newly named company retained a branch office at 25 York Street in Glasgow. The Lairds were not to become ship owners for some time but their interests were to influence the development of trade between both the north and the west coasts of Ireland from the Clyde (Chapter 5). Cattle and emigrants destined for the New World were the main live shipments to the Clyde while the shirt manufacturing industry in Derry supplemented inbound cargo.

The *Highland Chieftain* (1817) on the Clyde off Dumbarton.

[From a contemporary oil painting by William Daniell]

Laird (senior) was first registered as a ship owner in 1828 when he became part owner of the **Highland Chieftain**. She was running alongside the second steamer to be named **Comet** on the twice weekly service from Glasgow via the Crinan Canal to Fort William. This had been initiated by Henry Bell who had commissioned the first commercial sea-going steamship **Comet** in 1812. Laird was agent for this West Highland service from 1825 until 1832 and had an obvious interest in the vessel. In 1825 the purpose-built paddle steamer **Stirling** ran every other Wednesday from Glasgow to Inverness calling at Crinan, Luing, Easdale, Oban, Port Appin, Fort William, Corpach and Fort Augustus. A twice weekly departure had earlier been advertised until the **Comet** was transferred to Tobermory and Skye duties. Principal ownership of the **Highland Chieftain** was divided between Alexander Laird (senior), John Laird his brother, and Archibald McConnell. John Laird was a surgeon at Glasgow University and was obviously sufficiently in funds at that time to invest in the steamship, a fashionable means of committing spare cash within the upper classes throughout the early days of the steamship. The ownership of the **Highland Chieftain** passed to A MacEarchen in 1832 and Laird lost the agency role to him at the same time.

Londonderry was served by the Glasgow and Londonderry Steam Packet Company and it was this company that was eventually rebranded the Laird Line. The embryo company was started in 1814 by Lewis MacLellan who commenced a steamer service on the Firth of Clyde. This was gradually extended to Campbeltown in 1816, and included trips round Ailsa Craig in the following year. Then in 1820 MacLellan ran an excursion to the Giant's Causeway, repeated the following year with a visit also to Londonderry. In 1821 his steamer **Britannia** started a service between Glasgow and Belfast but it was not a success and in 1822 the ship was switched to a new weekly service between Glasgow and Londonderry. The vessel continued to serve Clyde ports and still offered trips to Londonderry via the Giant's Causeway as advertised in the Glasgow press:

> The steam boat **Britannia**, Captain Wyse, will sail from the Broomielaw Tuesday morning the 6th August at half past four o'clock for Greenock, Gourock, Largs, Ardrossan, Troon, Ayr and Campbeltown, from there proceed across the Channel to the Giant's Causeway, and other intermediate ports to Londonderry, and return to Glasgow on Friday the 9th August calling at the several Ports. As it is possible that this will be the last trip to the Giant's Causeway this season, the public would do well to embrace this favourable opportunity of visiting that romantic scenery, and other parts of the coast of Ireland. Glasgow, 30th July 1822.

The **Britannia** was joined by the **Argyle** in 1824 and maintained occasional trips to Dublin. A notice in the *Glasgow Herald* 4 November 1825 stated:

> In view of the advanced state of the season the Argyle and Britannia Steam Boats will cease plying to Londonderry during the winter season from and after this date and due notice will be given when they commence plying again.

The *Lairdsbank* (1893), originally the *Olive* – typical image of what was to come – heavily loaded with passengers on departure from Glasgow.

The north and west coast ships were always secondary in status to those on the Belfast and Clyde run, but were nevertheless important components of the trade between Scotland and Ireland. Laird Senior at this stage was not involved, but his son was destined to become the leader in the Glasgow to Londonderry trade.

With the onset of competition on selected routes, the little wooden paddle steamers were forced to work all year round in order to maintain their share of the market. Thus, for the first time the steamers were pitched against the ravages of the winter weather with the full stresses of wind, sea and tide. This at least gave full time employment to the seamen and firemen but under conditions that were at best hazardous and at worst downright dangerous. Passengers tended to travel on essential business only during these winter months, but parcels and other high value goods continued to be carried by the ships.

Thus far we see the inauguration of steamer services between the Clyde and Belfast and the Clyde and Londonderry. The difference between them was that the Londonderry service was unchallenged whereas that to Belfast was blossoming at such a rate that there was keen competition for the trade on offer. In 1826 James & George Burns were offering the **Fingal** with a cabin fare of 20/-, deck at 5/- from Greenock with a connecting river steamer to Glasgow, or from the Broomielaw direct if tides permitted; the **Ailsa Craig**, agented by R O Brown was sailing twice a week from the Broomielaw; and the 'Safety Steamer' **St Andrew**, agented by P & R Fleming, sailed twice weekly from the Broomielaw, 'touching' at Gourock. J & G Burns' other trade, which was to Liverpool, was joined in August 1826 by the New Clyde Shipping Company, which put their steamer **Enterprise** on the service in direct competition with the Burns brothers.

Nevertheless, the Belfast and Londonderry trades would ultimately become the core services operated by J & G Burns and the Laird Line. These services flourished for almost 150 years, before failing with the onset of containerisation and the roll-on roll-off ferry in the 1960s. Both companies were able to expand by acquiring competing companies, and both expanded their own routes to extend their business hinterland. Both were highly successful at their chosen trades, not only surviving but also expanding when other companies in the same trades were unable to compete. George Burns, of course had other interests as well, being instrumental in the foundation of what was to become the Cunard Steamship Company, and having earlier developed the West Highland trade, which was passed on to David Hutcheson and later David MacBrayne.

So successful were J & G Burns and the Laird Line that they were targeted as one of the first major acquisitions of the newly formed Coast Lines Limited in 1919 and 1920. Coast Lines had access to funds intended to expand its network available from the Royal Mail Group after the Elder Dempster Line had acquired the company earlier that year. Although the Burns and the Laird companies initially retained their own identity under Coast Lines ownership they merged in July 1922 to form Burns & Laird Lines Limited, adopting the Laird funnel colours and the Burns house-flag: a rampant gold lion clutching the globe on a blue ground (the Cunard house-flag was the same lion and globe on a red ground).

Burns & Laird Lines Limited flourished during much of the twentieth century. The magnificent motor ships **Royal Scotsman** and **Royal Ulsterman** were introduced to the Glasgow to Belfast overnight service in 1936 and a new ship entered the fleet as late as 1957 when the **Scottish Coast** first came up to the Broomielaw to serve on the Dublin service. The last purpose built cargo ship in the fleet was the cattle carrier **Lairdsglen** which was commissioned in 1954 and the very last ship built for the company was the drive-through ferry **Lion** commissioned in 1967. Coast Lines failed to acknowledge the roll-on roll-off concept until too late and was soon merged into the P&O Group where it quickly foundered to disappear from the waves once and for all.

The Burns brothers, notably George, and Alexander Allan Laird (senior) were important pioneer advocates of the paddle-driven steamship. They worked with many contemporaries to satisfy their vision of the steamer ousting the sailing ship from commercial employment in the coasting and cross channel routes. These included Henry Bell and his **Comet**, David Napier with his engineering skills, especially his revolutionary narrow-hulled steamships such as the **Rob Roy,** and many other great steamship entrepreneurs of the day. That the Burns and Laird family names would become inextricably linked one hundred years later by the formation of Burns & Laird Lines Limited would seem a strange concept to these men in the 1820s. This linkage, however, developed through parallel interests and a synergistic respect for each other's business. But come what may, the sailing ship was ultimately driven off the Irish Sea, and elsewhere, by people such as George Burns and Alexander Laird.

The cattle carrier **Lairdsglen** (1954) was the last purpose-built ship in the fleet.

[Nick Robins]

Lion (1967) was a belated attempt to enter the roll-on roll-off traffic on the Irish Sea with daylight services between Ardrossan and Belfast.

[Nick Robins]

Royal Ulsterman (1936) one of a series of motor ferries built for the Coast Lines group of companies, seen arriving at Glasgow on 30 July 1967.

[Nick Robins]

CHAPTER 2

WOODEN PADDLE STEAMERS BETWEEN GLASGOW, LONDONDERRY AND SLIGO

THE GLASGOW, DUBLIN, AND LONDONDERRY STEAM PACKET CO LTD.
This important company, which claims (on very strong evidence) to be the oldest steamship company in the world, was originated in 1814 two years after the launch of Bell's **Comet** by Mr Lewis MacLellan and others. Its history is a most varied one, the several firms of Alex. A Laird & Sons, Thos. Cameron & Co., and MacConnell & Laird, having become unified during its existence of nearly a century into the one large concern known throughout the kingdom as the 'Laird Line'. It has been the great pioneer of the steamship trade of the Clyde, not merely by reason of its long standing, but also because of the varied and extensive sphere of its operations.

From: *The History of Steam Navigation* by John Kennedy

The history of Burns & Laird Lines Limited starts in 1814 at the very beginning of commercial navigation by steamship. In that year Lewis MacLellan acquired the small wooden-hulled paddle steamer **Britannia** and set her to work on the Clyde. She was equipped with an American style beam engine. She was joined in the following year by a near sister, the **Waterloo**, although this ship had a conventional engine. Both the steamers were partly owned by Lewis MacLellan, Archibald MacTaggart of Cambeltown and others. Competition on the Clyde services became intense with new and larger ships being quickly introduced to the trade. MacLellan's response was to broaden his interests and by 1821 the **Britannia** was sailing as far as the Giant's Causeway and Londonderry, primarily as excursions. Several trial runs were also made to Belfast during that summer. The following year, the **Britannia** started a regular weekly service between Glasgow and Londonderry, calling at Campbeltown, with an inaugural sailing on 11 June (in those early days the small steamers were laid up safely in the winter months, secure from the ravages of wind and sea). That first sailing to Londonderry was the start of 144 years of continuous service on the Londonderry route by the so called 'Scotch' boat. The last vessel on the route was the **Lairds Loch** which closed the service in 1966.

Lairds Loch (1944) was built under licence in the Second World War and was designed for the Glasgow to Londonderry service. She is seen here at Glasgow on 23 August 1964 just two years before she closed the service.

[Nick Robins]

David McNeil describes the **Britannia**:

> The **Britannia** was a wooden steamer, 93 feet in length, and had a beam engine, a type popular in America but never fashionable in the British Isles. In her early days on the Derry - Glasgow route she ran in summer only and her Irish calls included Culmore, Quigley's Pont, Moville, Greencastle and Portrush. No provision seems to have been made for the carriage of deck or steerage passengers. Instead she had accommodation for first cabin and second cabin passengers who were charged 21s or 14s for a single journey. At Derry the agent for the **Britannia** was Mrs Cooley of the Marine Hotel from whose premises coaches set off for Enniskillen and Sligo immediately after the steamer's arrival.

In summer 1824 the **Waterloo** joined the **Britannia** on the Londonderry service, allowing two sailings per week. She was succeeded in 1824 by the **Argyle**, a steamer of the same pedigree; the **Argyle** dated from 1815. The ships called in addition at the Giant's Causeway, to be met by tender, and also at Coleraine.

The frequency of the service was temporarily increased to three times per week in June 1826 when the new steamer **Londonderry** arrived on station. She was a significantly larger ship than her predecessors with a length of 110 feet and a more powerful engine. This allowed her to reduce the journey time to 22 hours, instantly making her two running mates obsolescent. She had a stateroom for first class passengers, a cabin for second, and her steerage passengers were given shelter from the weather in what were called 'round houses' erected on deck. The **Londonderry** also had two small holds, fore and aft, with a deadweight capacity of 50 tons. She was joined by a similar ship, the **Clydesdale**, which was launched on 25 March 1826, so that the two sailings per week offered upgraded passage speed and accommodation. For the first time the service was continued throughout the winter months. The **Argyle** was disposed of once the **Clydesdale** was established in service. The ships still called at Campbeltown before setting out for the Irish coast.

Both the new ships were registered under one-ship companies, the Londonderry Steamboat Company and the Clydesdale Steamboat Company. This was a policy followed for subsequent ships built until the formation of the Glasgow & Londonderry Steam Packet Company in 1835.

In December 1827, demand was such that a third upgraded ship, the **Eclipse** was chartered for the Londonderry service. She was owned by James Dalzell, Robert Napier, James Stevenson, David Napier & James Moore and registered at Glasgow, having previously been operated between Greenock and Belfast. Her charter to Lewis MacLellan and his partners lasted through the winter, allowing the smaller **Britannia** to lay-up in dock.

The **Clydesdale** was lost when she caught fire off Rhins of Galloway on 15 May 1828. *Lloyd's List 20* May 1828, reported:

> Stranraer 16 May: The **Clydesdale** Steam Packet, [Captain] Turner, of and from Glasgow, to Belfast, took fire last night 14 miles to the NNW of Corswall Light, and was run on the Rocks close to the Light, where she lies a complete wreck. Crew and passengers saved.

It was over a year before new tonnage was available when the steamer **Foyle** was delivered in June 1829 to her registered owners, the one-ship Foyle Steamboat Company. She was the largest and fastest ship yet on the service, with a length of 122 feet and a passage time of just 20 hours. Overcrowding was a problem, particularly in the 1830s, when the **Foyle** could carry as many as 650 steerage passengers on deck, along with numerous live animals destined for market in Glasgow. Conditions for these passengers were little better than those provided for the animals, save that the cattle and sheep were fed and looked after by the herdsmen whilst on the ship.

The **Foyle** displaced the **Britannia** from the Londonderry route and then took up various duties running to Portrush, Coleraine and Newry. On 11 October 1829 the **Britannia** was leaking badly in a north westerly gale, and put into Donaghadee Harbour. The **Britannia** had been on passage from Warrenpoint to Glasgow under Captain Knox, who determined that the ship must discharge her cargo of grain, which had become wet and heavy, before she could continue her voyage. By 13 October the discharge was reported to be complete, although two thirds of the cargo had been damaged by water. At that stage the vessel sank alongside the stone harbour wall, while a north easterly gale forced a heavy swell into the harbour, smashing the hapless ship against the pier where she quickly became a total loss.

The **Foyle** took time off in the summer of 1830 to run several short cruises from Londonderry to Oban and Fort William, calling also at Staffa if the weather was kind. These were obviously successful as the West Highland cruises were advertised again in 1831 at a price of just 21/-.

Another new larger and faster wooden-hulled paddle steamer, the **Saint Columb,** joined the Londonderry service in August 1834 (Londonderry Cathedral is dedicated to St Columb). Sailings were made from either port on Tuesdays, Thursdays and Saturdays, subject to tidal conditions. The **Saint Columb** made the trip from Glasgow on her maiden voyage, quay to quay, in eighteen hours. As with the **Foyle** and her predecessors, the new ship was registered under a one-ship company. From 1829 onwards, Glasgow Fair Week was celebrated by an excursion sailing via Londonderry to the West Highlands. This was initially operated by **Foyle**, but from 1835 onwards this duty fell to **Saint Columb** when the excursion was extended to include a visit to Portree.

The **Saint Columb** allowed the **Londonderry** to be sold in January 1835; the core service was then maintained by the **Foyle** and **Saint Columb**. A rival ship plied on the route from Glasgow in September 1835 and steerage class fares fell to just 1/- (5p in modern parlance). The rival was the former West Highland steamer **Glen Albyn** owned by the Glen Albyn Steamboat Company of Tobermory. She remained on station at Londonderry throughout the winter but quickly gave up the competition in the late summer of 1836 before reverting to West Highland duties. Fares duly regained their commercial value and calm was again restored. In truth there was room for the competitor, as ships were grossly overcrowded with steerage passengers herded on deck along with cattle, sheep and pigs. Many of these passengers made the crossing from Ireland with their eyes set on a passage to America from the Clyde. However, if they had neither an onward ticket nor the resources to buy a ticket for America they were put back on the boat and sent home the next night; at least the return trip was without animals penned on deck. Others made the passage from Ireland as reapers or seasonal workers in other trades in summer, returning to their families in the autumn. But whatever their purpose the steerage passenger had an uncomfortable journey exposed to the elements.

The **Saint Columb** and **Foyle** were transferred to the newly-formed Glasgow & Londonderry Steam Packet Company in December 1835. This company was formed by amalgamation of the two one-ship companies, the Foyle Steamboat Company and the Saint Columb Steamboat Company. Messrs. Thomas Cameron & Company were appointed Glasgow agent for the new company, having already been associated with the Saint Columb Steamboat Company for some time. MacLellan's venture into the Irish trade had indeed blossomed and the fruit of this had become an important shipping company based in Glasgow.

The Glasgow & Londonderry Steam Packet Company took delivery of its first ship in September 1836, when the wooden-hulled paddle steamer **Rover** was delivered. She could carry 60 saloon passengers, most of whom were accommodated in the grand cabin, a large public room for a ship of such modest dimensions. Her master was Captain Wyse, the **Foyle** was under the command of Captain Turnbull and the **Saint Columb**, Captain A Coulter. The **Rover** was known to have carried as many as 1,500 steerage passengers bound for Glasgow. This ship provided much needed support for the two departures from either port on Monday afternoon and Friday afternoon. Competition again reared up when the North British Steam Navigation Company put the **Antelope** on the route from the Broomielaw in 1837, with Hugh Price as its Glasgow agent. Owners of the North British company were none other than James Martin and James & George Burns. This was a blatant attempt by the Burns brothers to break into the Londonderry trade. However, the **Antelope** remained on station only until early 1839 when depressed rates on the service forced her to be retrenched to Belfast-Glasgow duties.

The Glasgow & Londonderry company extended its sailings to Sligo in 1841. The **Saint Columb** arrived at that port for the first time on 29 September to commence a fortnightly service from Glasgow. She was replaced by the slightly faster **Rover** in the following year allowing a 22 hour voyage to the west coast port, with calls at Campbeltown and Londonderry. This new service was made possible by the addition of the new steamer **Londonderry**, second of that name, which entered service in August 1841. This **Londonderry** was the last wooden-hulled vessel to be ordered by the company. She was the biggest ship in the fleet so far, with a length of 157 feet and of 513 tons gross. The Sligo service was so successful that a Liverpool to Sligo service was commenced in 1843, generally on a monthly basis. Again both services developed to satisfy the demand for live cattle, sheep and pigs for slaughter and emigrants destined for the New World. However, conditions for steerage class were little better than they had been in the 1830s. Another new service was the weekend return leaving Londonderry on Saturday mornings and returning from Belfast on Monday afternoons operated by the **Londonderry**, **Saint Columb** or **Rover**.

On 31 March 1844 the **Saint Columb** struck a submerged rock in the Sound of Sanda and stranded. The master put his passengers onto a nearby rock from which they were transferred to Sanda island. The Campbeltown steamer **St Kiaran**, coming along the next day, not only rescued the passengers but also succeeded in towing the **Saint Columb** off her perch before the damage had become too extensive. The **Saint Columb** was able to put back to Glasgow under her own steam. Later in the same year, the **Londonderry** and the **Dolbadarn Castle** of Caernarvon, on passage from that port to Glasgow, were in collision in the morning of 10 September. The accident happened near the Toward Light in the Firth of Clyde; the **Dolbadarn Castle** sank rapidly while the **Londonderry** remained watertight, her boats gathering survivors from the stricken ship.

A new iron-hulled paddle steamer, the **Rambler**, was launched on 26 March 1845 and commissioned shortly afterwards. The order for this new ship had allowed the **Foyle** to be sold during the previous autumn. **Foyle** had several subsequent owners and provided many more years of useful service – she was eventually converted into a sailing ship. The **Rambler**, which was primarily intended to support the burgeoning services from Sligo, very nearly came to grief in her early career as reported in *Lloyd's List* 27 May 1846:

> Liverpool 26 May. The [steamship] **Sea Nymph**, for Newry, and the **Rambler**, from Sligo, with a large number of emigrants for embarkation hence to New York, ran foul of each other last night, off the Rock Lighthouse; the **Rambler** was run on shore to prevent her sinking, and fills with the tide; the **Sea Nymph** returned into dock with considerable damage, and leaky. Several of the **Rambler**'s passengers were killed or hurt by the collision, and others drowned by jumping overboard.

The following day the **Rambler** was refloated in the evening and taken safely into dock. She was quickly repaired and refurbished although no attention was made to giving the ship any form of watertight bulkheads, which would have saved the ship from sinking in the first place. Unhappily, the story did not end there. Repaired and ready for service once again, she sailed from Liverpool for Sligo on 14 June, *Lloyd's List* 17 June 1846:

> Belfast 15 June. The **Rambler**, [Captain] M'Alister, from Liverpool to Sligo, ran on the Maiden Rocks during a dense fog yesterday morning, and filled; Crew and passengers saved; a portion of the cargo landed in a partly damaged state.

It quickly became apparent that the **Rambler** was likely to become a wreck and she was still on the rocks in early July when parts of the engine were removed to lighten ship. Inevitably the sea started to break up the hull and she became a total loss. Interestingly her name, **Rambler**, as in rose, and the first of the horticultural names to be adopted by the company, was never used again. The remaining parts of the engine were removed, the engine reassembled and stored, ready to be installed in the new steamer **Thistle** in 1848 (Chapter 3).

Shortly afterwards, and almost two years after the stranding of the **Saint Columb** near Sanda, the **Londonderry** repeated the accident. She had been on passage from Londonderry to the Clyde in foggy conditions under the command of Captain Wyse when, on the evening of 20 August 1846, the **Londonderry** struck the rocks at Keill, near Campbeltown, and stranded. The passengers and crew were all taken safely ashore but the ship remained where she was for some days before the tide allowed her release. The weather had been kind and the **Londonderry** was surprisingly little damaged.

Alexander Laird and his namesake son had also been busy. Now based at Glasgow, they had become arch rivals of Thomas Cameron and his son John Cameron, to the extent that there was absolutely no collusion between the two agents. The Lairds had briefly been agent at Greenock for the St George Steam Packet Company and then continued when the terminal moved to Glasgow after a short break in the service. Laird was also agent for the Mersey & Clyde Steam Navigation Company from 1823, with a new office at 25 York Street, Glasgow, under the title Alex. Laird & Co. *The History of Steam Navigation* records that:

> The pioneer steamer of the new service was the **Henry Bell**, built by Mr. Thomas Wilson, a celebrated Liverpool shipbuilder, in 1823. She was considered a very smart craft in those days, was fitted with two engines of 30 horse power each, and carried about 120 tons all told, on a draft of about 8 feet. She continued on the Glasgow and Liverpool station until 1831, when she was purchased by Messrs. James Little & Co, for their Glasgow and Newry trade.
>
> The original intention of the proprietors was that the **Henry Bell** should sail to and from Glasgow, but it was found there was not sufficient water in the Clyde to enable this to be done with regularity, and Greenock was, consequently, made the port of arrival and departure. The deck fare by this steamer was 6/-. per passenger, the steerage fare by the mail packets (**Majestic** and **City of Glasgow**) being 21/-.
>
> The late Mr Alex. A Laird commenced his apprenticeship under his father in 1824, and the same year a second vessel, the **James Watt**, was placed on the Glasgow and Liverpool station. She was slightly larger, and had engines of greater power than the **Henry Bell.**
>
> The following year Messrs. Laird established a fortnightly service between Glasgow and Inverness; the steam packet employed was the **Stirling**, which made her first voyage on the 11th May 1825, and continued to sail thereafter on alternate Wednesdays from Glasgow and Inverness. Fortnightly sailings proving insufficient for the traffic, the sailings were increased to weekly on and from the 20th September, 1826. During this year the **William Huskisson** was added to the Liverpool and Glasgow service, and sailings were maintained three times per week from each port. The **William Huskisson** was a very much larger vessel than either of her predecessors, her deadweight capacity being 350 tons, and her engines 120 horse power.

An advertisement in the *Glasgow Herald* 26 May 1828, read:

> FOR LIVERPOOL, the Royal Mail and War Office Steam packets
> **Henry Bell** – Captain Spiers
> **James Watt** – Captain Johnston
> **William Huskisson** – Captain Lindsay
> Are intended to sail, with or without pilots, viz. FROM GREENOCK
> **Henry Bell** at 3 pm on Tuesday, May 27
> **William Huskisson** at 3 pm on Saturday May 31.
>
> And from the Broomielaw, **James Watt** at 11 am, on Wednesday May 28, calling at PORT-PATRICK and the ISLE OF MAN, to land and receive passengers.
>
> The goods by the former vessels are forwarded to Glasgow, in lighters, by the first tide after its arrival at Greenock, and at Glasgow, at 8 o'clock the night previous to its sailing, at the company's expense. There are round houses on deck for the accommodation of deck passengers.
>
> For freight or passage, apply at the company's office, at 79 Millar Street. ALEX. LAIRD & COMPANY.
>
> The **Clarence** steam-boat sails from the Broomielaw on Tuesday and Saturday, at 12 o'clock noon, with the passengers and parcel bags.

History of Steam Navigation again:

> For the Campbeltown and Londonderry trade the steam packets **Clydesdale** and **Londonderry** were built, and were advertised to sail from Glasgow to both ports, with goods and passengers, every Tuesday, Thursday and Saturday. In addition to these sailings, the **Maid of Islay** was dispatched every Tuesday morning from the Bromielaw to Stranraer and Islay. Messrs. Laird's connection with Dublin dates also from this year, the pioneer steamer being the **Town of Drogheda,** which sailed on her first voyage from Greenock to Dublin on Monday, 7th June, 1826. The new steam packet **Solway** was added to the Liverpool and Greenock fleet in 1828, and the sailings increased to four per week from each port. The steam packet **Clarence** acted as tender, and sailed from the Bromielaw at noon on the sailing dates of the Liverpool steam packets from Greenock.

The Mersey & Clyde Steam Navigation Company was absorbed by the newly formed Glasgow & Liverpool Shipping Company in 1831. Laird lost this agency at that time to Robert Gilchrist & Company.

In addition Alexander Laird also had interests in the West Highland trade between 1825 and 1832. This was as agent for the line of West Highland steamers developed from Henry Bell's **Comet** in 1812. **Comet** was succeeded by **Stirling Castle** and **Highland Chieftain** in 1820, allowing a twice weekly service, and joined by a second **Comet** and the **Highlander** in the following year. The ships principally served Fort William from Glasgow, and from 1824 they continued on to Inverness with time saved by using the newly-completed Crinan Canal out of the Clyde estuary. There was also a secondary service to Tobermory from 1822 onwards. In 1834 Laird senior formed a partnership which became Messrs. Laird & Bruce at Glasgow, but that was dissolved in 1835 when Alexander Laird & Sons opened an office at Glasgow. Thus an advertisement in the *Glasgow Herald* 20 October 1834, read:

> THE SAINT GEORGE STEAM PACKET COMPANY'S VESSELS
> **Saint George** – Captain Oman
> **Saint David** – Captain Moffat
> Are intended to sail from The Broomielaw, with goods and passengers, the **Saint David** for NEWRY and DUBLIN, this day (Monday) till 12 o'clock, and sail at 1 precisely. The **Saint George**, for DUBLIN direct, on Wednesday 22nd inst., at 2 pm.
>
> All these vessels are intended to sail precisely at the hours advertised, Goods for shipment must be alongside at least an hour before sailing. For freight or passage apply in Greenock to ALEX. LAIRD & SONS, or here to LAIRD & BRUCE, 69 Oswald Street, Glasgow.

In January 1836 the vessels advertised were the **Jupiter**, built in 1835 and under the command of Captain Oman, and the **Saint George**, now under the command of Captain Moffat. The Glasgow agent was then Alex. Laird & Sons, 12 York Street. The ships now called at Gourock and sailed direct to Dublin, with a sailing in either direction approximately every eight days. From 1836 onwards the rival and newly formed Dublin and Glasgow Steam Packet Company (see Chapter 4) ran the **Arab** – her master was Captain James Oman formerly of the rival company's **Saint George** – on the same route, soon joined by the **Mercury**, with twice weekly sailings. The Saint George company had been in financial trouble for some time, having overstretched itself in the North Sea as well as the Irish Sea and withdrew from the Glasgow to Dublin route early in 1839; Laird again lost an agency, the new Dublin & Glasgow company having appointed Lewis Potter as its Glasgow agent. However, in due course Laird was to share the agency, becoming responsible for the ships on the Cork service via Dublin, owned by the St George company, notably the **Ocean** and the **Minerva**. These ships ran in conjunction with those of the Dublin & Glasgow Steam Packet Company.

In 1844 Laird & Sons (then at 101 Union Street, Glasgow) was appointed agent for the Dundalk Steam Packet Company, the first steamer from Glasgow, the **Finn MacCoul,** sailing on the 30 November 1844. Thus, although Alexander Laird & Sons was not in any way associated with the Londonderry to Glasgow route at this time, it was well equipped with the right experience, should opportunity arise.

New Directory of the City of Londonderry 1839

The Directory reports 'The markets of Londonderry are, generally speaking good; the quays are commodious; the trade export, import and retail increasing; the town extending and improving. There are also six steam-boats attached to the port, by which passengers, as well as goods are conveyed to Liverpool, Glasgow and Campbeltown. The extensive shipbuilding establishment of Mr Coppin also deserves particular notice. Here a patent slip had been constructed for vessels of small burden; but by this gentleman's improvements, it is being enlarged so as to admit of vessels from five tons to six hundred tons register. There is also an extensive foundry, and steam boiler manufactory; and steam engines for vessels of every class are constructed on the premises.'

The Directory lists eleven master mariners; eight ship owners: S and J Alexander, Danie Baird, James Corscadden, John Cooke, John Kelso, William McCorkell, R and W F McIntire and John Munn; and four shipbrokers. There were 21 haberdashers and linen drapers, three flax spinners and 13 woollen drapers, suggesting a burgeoning linen and textile industry that would later lead to the famous shirt making industry in the city. There were numerous other diverse trades listed, several pages of publicans such that the Directory collectively gives the impression of a thriving port and market.

CHAPTER 3

ENTER MR ALEXANDER A LAIRD JUNIOR

The new flagship of the Glasgow & Londonderry Steam Packet Company was the **Shamrock**, the second iron-hulled paddle steamer commissioned by the Glasgow & Londonderry company. She was launched on 15 May 1847 from the yard of Caird & Company at Greenock and was fitted with the new hybrid 2 cylinder lever engine, precursor to the side lever engine.

Conditions on the ships in the Irish Sea remained appalling, especially in the winter months in steerage. Records show that 1,778 passengers, almost entirely steerage and almost all emigrants, were, for example, packed aboard the 513 gross ton steamer **Londonderry**, when she sailed for Glasgow on 11 August 1847, albeit in summer and calm weather.

Emigration was at its highest in the late 1840s during the Great Famine. Between 1845 and 1852 some 2 million people fled Ireland in search of a better life in the New World, while a further 1 million people died at home. These awful statistics, aided by the 'Synthetic Famine' in 1850 and 1851 when Westminster did little to exacerbate the hardship, enhanced steerage passenger carrying out of Ireland, while the carriage of cattle and sheep continued in return for inadequate volumes of grain. The longer term impact of the famine was a hugely reduced population in Ireland and a growing animosity towards both absentee English landlords, who had evicted their impoverished tenant farmers, and the remote and disinterested rulers at Westminster.

By December 1847, the core service to Londonderry was operated by the **Londonderry** and the **Rover**, the latter ship also calling at Portrush (weather permitting). A new service from Troon to Portrush and Londonderry was commenced at the beginning of the month by the **Saint Columb**, departing Troon every Wednesday and Saturday afternoon, and returning from Londonderry every Monday and Thursday. A through train to and from Glasgow was operated by Ayrshire Railway; the fares from Glasgow to Londonderry with train connection were first class 14/-, second class 12/6d and steerage 3/-. The Sligo service was operated by the **Shamrock**, with two sailings a month also to Liverpool. Interestingly the Liverpool agent was T Martin, Burns & Company. There was also a rival steamer, the **John Munn,** which sailed on Mondays from Glasgow to Londonderry calling at Portrush rather than 'calling off', and returning from Londonderry on Thursdays. She was owned by John Munn, junior, of Foyle Street, Londonderry. This service lasted for just over a year.

By mid-year 1848 the twice weekly Londonderry service was operated by **Shamrock**, the **Saint Columb** was still based at Troon and the **Londonderry** was running Sligo to Liverpool and Glasgow.

Robert Napier & Company at Govan launched the next ship for the company on 2 September 1848 with the name **Thistle**. She was another iron-hulled paddle steamer of slightly larger dimensions than the **Shamrock**. Her engine was second hand, having been salvaged from the **Rambler** when she was lost in 1846. The **Thistle** commenced service to Londonderry in November, working initially alongside the **Rover**.

The **Londonderry** got into trouble in August 1848 as reported in *Lloyd's List*:

> Londonderry 3 August. The **Londonderry**, from Liverpool to Sligo, and the [steamship] **Connaught Ranger**, from Sligo to London, have put in here, the former with the upper part of her starboard bow stove in, and the latter with stem started, and making water, having been in contact this morning, off Innistrahull.

The *Liverpool Mercury* followed on 8 August 1848 with the announcement:

> The ss **Londonderry** will not sail for Sligo today as formerly advertised, in consequence of having met with a slight accident.

At the end of October 1848, the Troon service was closed down and the **Saint Columb** was placed on the Glasgow to Portrush service. By December 1848, the **Shamrock** and **Rover** were maintaining the basic Londonderry service, the **Rover** was running to Sligo alongside the **Londonderry**, and the **Saint Columb** was serving between Glasgow and Portrush, while the **Thistle** was off service. The **Londonderry** sailed from Sligo for Liverpool on 1 December 1848 in bad weather with three saloon passengers aboard and 174 steerage class passengers herded into the round houses. She ran into a fierce gale and her master, Captain Johnston, considered it prudent to put his steerage passengers in the forecastle for fear they would otherwise be washed overboard. McNeill wrote:

> The weather was so bad that the steerage passengers were put in the hold and battened down to keep them dry. When the hold was opened next morning it was found that 72 passengers had died from lack of air. There was a general outcry and it was pointed out that the Black Hole of Calcutta was more capacious than the hold of

the **Londonderry** (23 feet x 8 feet x 6½ feet). Nevertheless the general feeling of those in authority was that the emigrants were entirely to blame since they had overcrowded the steamer.

Many of the cattle penned on deck were lost. The ship arrived in a poor state at Londonderry on 3 December, having been at sea for over two days. A correspondent for the *Glasgow Herald* reported on 8 December 1848:

> I write to inform you that one of the most awful occurrences that has ever taken place at sea, has this day been brought to light at this port [Londonderry]. The steamer **Londonderry**, Johnston, master, bound from Sligo to Liverpool, with 174 passengers, left the former port on Friday evening, and as it blew hard, shortly afterwards stopped, for a couple of hours, at the pool, near the entrance to that harbour.
>
> The following is the substance of a witness examined at the inquest. After remaining about a couple of hours at the pool, the vessel proceeded on her passage. Went below about 1 o'clock on Saturday morning, by command of the captain. Was the last man who went below. After remaining about half-an-hour below, began to feel a great sense of suffocation, occasioned by the large number of steerage passengers in the forecastle, without any ventilation whatever, which was greatly increased by the circumstance of one of the passengers having accidently lighted a box of lucifer matches. With the greatest difficulty, from weakness, and the number of people lying on and about him, he made his way to a place where he got a little air, which revived him sufficiently to enable him to force his way on deck. He then found that the companion had been completely closed by a tarpaulin. On making his way to a seaman, he was met by curses and ordered below, with threats that he would otherwise be thrown overboard. Had great difficulty in getting the mate induced to go down, when they found a large proportion of the passengers suffocated. Witness then assisted them to get out all the passengers showing signs of life.
>
> The vessel arrived at this port around 9 o'clock this morning, having reached the lough yesterday evening, but was unable to proceed up the lough. I have omitted to mention, in this hurried statement, that the witness stated also that the captain was not to be seen by him for three or four hours after the circumstances above alluded to were discovered. Since the arrival of the vessel, the steerage passengers, captain and crew have been taken into custody. I regret to state that, since the above was written, it has been discovered that seventy two of the passengers have perished from suffocation. I cannot conclude this account without remarking that it is disgraceful that in this age of enlightenment the arrangements for steerage passengers in steamers generally are so defective in ventilators, and also that individuals should be placed in charge of these vessels displaying such gross ignorance of the necessity of ventilation as has been shown in this instance…

The ship was impounded and the **Thistle** was brought back into service to cover, and later in the month the chartered Burns' steamer **Aurora** took over. However, the **Londonderry** and her crew were quickly released and returned to their Liverpool to Sligo duties. Indeed on 9 December 1848 *Lloyd's List* reported: 'The **Mary Ellen**, which arrived here from Montreal, lost her jib-boom and cutwater last night by being run foul of by the **Londonderry**'. Thus by the end of the first week of January the chartered **Aurora** was returned to Burns' Liverpool and Glasgow duties.

For several months in 1849 the **Thistle** and **Londonderry** were chartered to the Fleetwood-based North Lancashire Steam Navigation Company (a subsidiary of Kemp & Company of Fleetwood) for its short-lived Fleetwood to Londonderry service. The service had commenced in January 1849, using the Fleetwood steamers **Prince of Wales** and **Princess Alice**, but was closed later in the year due to lack of patronage.

The next new ship ordered by the Glasgow and Londonderry company never actually sailed for them. This was the **Laurel**, which was laid down at Caird & Company's yard and launched on 27 May 1850, but completed for the Burns owned Glasgow & Liverpool Steam Shipping Company as **Laurel**. In the following year, the new ship **Rose** was delivered to the company, allowing the faithful **Saint Columb** to be sold out of the fleet. The **Rose** was launched on 10 September 1850 and was ready to take up service to Londonderry on 28 January 1851. She had a side lever engine and was fitted with the newly developed feathering paddle wheels, a system designed to provide greater efficiency than the fixed float type paddle wheel. However, the newfangled floats were so difficult to maintain, even described as dangerous, that they were replaced by fixed floats in her second year of service. Nevertheless, she was a fast ship and could make the passage from Greenock to Londonderry non-stop in 10 hours.

The smaller ships operating on the Sligo service occasionally extended the service to call at Ballina and Westport, but for cargo and livestock only. These were not regular calls but rather based on inducement, the first departure from these small ports was in 1851.

On 4 February 1851, the **Thistle**, on a voyage from Glasgow to Londonderry, was in collision with the brigantine **Laurel**, off Lamlash. The **Laurel** which was arriving in the Clyde from Demerara, sank with all her crew. Wreckage was driven ashore on the Cumbraes, and along the beach near Largs. The **Thistle** had to return to Glasgow with a damaged bow.

Three new and significant initiatives took place in 1851. In April 1851 the North Lancashire Steam Navigation Company attempted to renew its Troon sailings by calling at that port on their Fleetwood to Londonderry service. This was little cause for concern for the Directors of the Glasgow & Londonderry company until the following year, when the North Lancashire company placed a steamer on a direct Glasgow to Portrush and Londonderry service, primarily targeted at livestock carrying. The ships were owned by Kemp & Company: **Royal Consort, Princess Alice** and **Fenella**, and all were modern paddle steamers. They were joined by the **Cambria** and **Prince of Wales** in the following year. The service thrived and in 1853 was taken over by MacConnell & Laird under the banner City of Glasgow Steam Packet Company. The cabin fare was offered at 12/- and steerage at 3/-. Departures from Glasgow were on Tuesday and Friday returning on Monday and Wednesday.

Prince of Wales (1842) was the only steamer McConnell & Laird registered in their names in 1853, formerly of the fleet of Kemp & Company of Fleetwood.

[From a contemporary oil painting, Fleetwood Museum]

Archibald MacConnell was also a director of the Glasgow & Londonderry Steam Packet Company. He was able to do a deal with Cameron & Company, managers of the Glasgow & Londonderry company, such that the two firms combined to provide an integrated service to Londonderry, the new vessels sailing on days when the Glasgow & Londonderry company did not. The paddle steamers were gradually phased out. The new screw steamer *Emerald* arrived on station in 1855; her registered owners were Robert Armour, Patrick J Mills, William Whyte, Alexander A Laird & James R Napier. The *Emerald* worked with the *Dundalk*, a modern steamer chartered from the Dundalk Steam Packet Company. In 1857 chartering ceased when the new steamer *Eagle* was delivered; her registered owner was Alexander A Laird, the son of the former shipbroker and agent who had first started in Greenock in the early part of the nineteenth century. Alexander Laird had finally arrived on the Londonderry route, for which he was to become best remembered.

Neither the *Emerald* nor the *Eagle* was to remain in Laird's ownership for long and both ships were lost in 1859. Laird sold the *Emerald* in 1857 and under the ownership of the Ayr Steamship Company (Chapter 8), she was wrecked on Chicken Rock (Carrick ny Kirkey in Manx), Isle of Man, on 24 August. *Lloyd's List* 29 November 1859, then reported:

> Greenock 26 November. The *Eagle* (s), from Glasgow to Londonderry, came in collision last night off Lamlash, with the *Pladda* (barque), from Quebec to the Clyde, and sunk; second mate and several passengers reported drowned. The *Pladda* is believed not to have received much damage.

The *Eagle* was still managed by MacConnell & Laird and a replacement was immediately ordered. This was the *Falcon*, another iron screw steamer registered under the ownership of Alexander A Laird at Glasgow. Laird was now clearly a part of the establishment and any animosity that had previously been felt between Laird and Cameron was put behind them.

The second new initiative was another new service to Londonderry which started in 1851, this time very much with the approval of Messrs. Cameron & Company and the Directors of the Glasgow & Londonderry Steam Packet Company. This was a collaborative project with the Midland Railway. It provided an extension of some of the railway company's sailings from Morecambe to Belfast to continue on to Londonderry by deploying spare tonnage from the Glasgow & Londonderry company's fleet. It was started in December 1851 by Cameron's paddle steamer *Rover* and then left to the *Rose* and *Laurel*. However, a twice weekly service was found to be unwarranted and it was left mainly to *Rose* to maintain a weekly departure from both Morecambe (then called Poulton-le-Sands) and Londonderry. Various other ships were used at Morecambe, but the service remained poorly patronised and Cameron & Company elected to withdraw from the route in January 1853. It had not been without its problems and the confined approach to the Stone Pier at Morecambe caused a number of mishaps.

Mr Alexander A Laird 1811-1892

The Midland Railway then chartered ships for the route. However, in 1854 the Glasgow & Londonderry company's paddle steamer **Laurel** and the screw steamer **Arbutus** were 'sold' to the railway company and registered at Lancaster in the names of trustees of the 'Little' North Western Railway, ready to take up the service.

The **Arbutus** was a screw steamer built with the Glasgow to Londonderry service in mind, but completed for the North Western Railway and registered at Lancaster in 1854. She was the first ship ordered by Cameron from outside the Clyde, being completed by Thomas Toward at Newcastle with engines built by Hawthorn & Company. Under railway ownership the steamers continued to be managed by Cameron & Company, a relationship that was clearly satisfactory to both parties as it remained in place for the next eleven years.

On 10 August 1854, the **Laurel**, under Captain Selly, took the ground on Mort Bank in Morecambe Bay on arrival from Belfast. She was later floated off as the tide rose, but the engine room and cabin promptly flooded and she had to be beached, later going on the Patent Slip for permanent repairs. In early 1855, the Liverpool Dock Registers show the **Laurel** working briefly from Clarence Dock to Belfast, relieving Langtry's **Waterloo**.

The third new initiative that started in 1851 was destined to involve the Glasgow & Londonderry company only at a later date. The Dublin & Glasgow Sailing & Steam Packet Company of Dublin ran a successful service between its namesake ports but was concerned that an opposition company could start to erode its business. In order to ward off competitors it established a new company, the Glasgow & Dublin Steam Packet Company, to run an independent competitive weekly cargo service with the little screw steamer **Northman**. The **Northman** had been built for a Kirkwall company in 1847 and was bought in 1851 and registered under the ownership of none other than Alexander A Laird, who was Glasgow agent for the Dublin company. By 1854, the parent Dublin company no longer recognised a threat to its own service and determined to sell the **Northman** and close the quasi-competitive service down. However, Laird had built up a considerable business between the two ports and ordered a new steamer, the **Irishman**, to carry on the service in his own name. This ship was launched in May 1854 and was ready for service shortly afterwards. Her machinery comprised a pair of steeple engines. Her funnel was painted white with a black top in the style of the St George Steam Packet Company. Her registered owner was Alexander A Laird, the company still trading under the name Glasgow & Dublin Steam Packet Company. Clearly Mr Laird was becoming a rising star.

On 11 March 1852 the **Rover** put into Killybegs, on passage from Londonderry to Glasgow, where she struck on a shoal in the harbour. However, she was able to proceed in a leaky state to Glasgow, where she was repaired. But worse was to follow, as reported in *Lloyd's List*, 11 December 1852:

> Portrush, 8 December. The **Rover** (s), bound to Glasgow, is on shore near Dimscrenick Castle, with bows stove, full of water, and with cargo washing out, having been in contact with the **Princess Alice** (s), from Glasgow, which has lost fore sponson. The **Rover** is expected to be a total wreck; crew and passengers saved.

The fleet of Cameron's Glasgow & Londonderry Steam Packet Company continued to upgrade. The **Rose** was joined in 1854 by near sister **Myrtle**, save that the latter had three masts instead of two. She was also slightly smaller, but a little faster, than her consort. The **Myrtle** was distinguished for two reasons. First, she was the last paddle steamer to be commissioned by the Glasgow & Londonderry Steam Packet Company and, secondly, because of the short tenure she had with that company. Her first voyage to Londonderry was in January 1853 and her last was in August. *Lloyd's List* 7 August 1854 reported:

> Campbeltown 5 August. The **Myrtle** (s), from the Clyde to Londonderry, got on the rocks in the Sound of Sanda, yesterday morning, has thrown overboard part of cargo (iron), has received considerable damage, and fills with the tide.

And on 31 August 1854:

> Greenock 29 August. The **Myrtle** (s), from the Clyde to Londonderry, which went ashore in the Sound of Sanda on 4 August, has gone to pieces. Her engines, with the exception of the part broken by the concussion, have been got out of her and taken to Glasgow.

The ship had struck the Barron Rock on Sanda Island, near Campbeltown, in hazy weather. The vessel and her cargo were insured for a total £14,000. No lives were lost in the incident. The *Glasgow Herald* reported on her salvage on 6 August 1865:

> The wreck of the iron steamer **Myrtle**, 212 tons register, as she at present lies under water in the Sound of Sanda, was sold by Messrs. Laughland & Brown on Tuesday for £5. The portion of the working parts of the engine and materials saved from the wreck brought about £400.

Accidents continued to happen. The **Shamrock**, under Captain Stewart, sailed from Glasgow for Sligo on the evening of 1 August 1855. Early on the following morning she struck rocks in the Sound of Sanda and quickly filled with water.

Nobody was hurt in the incident. On 18 September it was reported that the ship had been patched up and refloated, such that she could be taken to safe haven at Campbeltown for permanent repairs to be carried out.

The replacement for the **Myrtle** was the magnificent **Garland**, the first iron-hulled screw steamer built for the company. She was equipped with a horizontal type steam engine, built by A & J Inglis, to a design that was intended to reduce the metacentric height of the machinery. The ship was launched from the yard of Robert Napier & Sons at Govan on 4 August 1855, and was ready to start on the Londonderry service a few weeks later.

In 1857 a small wooden lighter was completed for John Cameron & Lewis MacLellan and named **Bee**. She was used for the next ten years on feeder work including the east coast of Scotland via the Forth & Clyde Canal.

The **Thistle** ran down and sank the wooden-hulled steam tug **Duke of Northumberland** off Cumbrae in the early morning of 17 May 1858, when inbound from Londonderry to Glasgow. The accident happened in a dense fog. The crew of the tug was able to launch their boat and scramble safely aboard before the tug sank. The master of the **Thistle** did not know they were safe and stayed in the area for some time looking for survivors; the crew of the tug landed at Millport that afternoon. The tug was owned by the Clyde Shipping Company and registered in the name of John Kidston.

The **Thistle** was again in trouble later in the year. Lloyd's List reported on 11 December 1858:

> Sligo 10 December. The **Thistle** (s), from Glasgow to this port, went ashore last night near Raghly Point, and a total loss is apprehended, as it is blowing strong.

Within two days it was reported that nearly all the cargo, excepting the sugar which was damaged, had been landed, and that the vessel lay on the rocks in an exposed position. Within the week the vessel was breaking up, the wind having shifted to the south west. What remained of the wreck was salvaged the following year.

The Board of Trade held a preliminary enquiry before the Inspecting Lieutenant of Coastguard, Sligo, which determined: The loss of this vessel was occasioned by the Master imagining himself closer to the Rock Light than was really the case, a conclusion favoured by the darkness of the night and a peculiarity of the atmosphere, which caused the light to assume a very 'glary' appearance, making a correct judgement of distance by its aid a matter of extreme difficulty. This circumstance was confirmed by other masters. There was no neglect or inattention by the Master, but the casualty arose from a want of judgement as to the distance of the light.

The **Rose** ran ashore on Copeland Island, near Donaghadee, on 27 July 1858, on passage from Morecambe to Belfast. Her fore hold quickly filled with water but passengers and crew were landed safely, the weather being fine with a slight north easterly wind. By the following morning she was largely under water, only part of her cargo being salvaged. However, by mid-August the ship had been raised and taken into Donaghadee Harbour. The **Rose** was then repaired, refurbished and put back into service. She was again in the news when on 13 December 1858 she broke her intermediate paddle shaft and had to be towed into Londonderry by the steamer **Culdaff Bay**.

A new **Thistle** was launched on 2 July 1859 and was equipped with American style machinery, as reported in the Glasgow Herald 4 July 1859:

> Launch at Port Glasgow: On Saturday last, a beautiful steamer, which now bears the name of the **Thistle**, was launched from the building yard of Mr Laurence Hill, of Port Glasgow. She is the property of the Glasgow & Londonderry Steam Packet Company, and will replace the late steamer of the same name so long and favourably known in the Irish trade. A select party of ladies and gentlemen proceeded from Glasgow by the 11 o'clock train and beguiled the time till 1 o'clock by surveying the vessel as she stood on the stocks, and examining other matters of interest in Mr Hill's yard. At the hour mentioned the dog-shores were knocked away, and the new steamer glided into the Clyde, the ceremony of christening having been very gracefully performed by Miss Annie M'Gregor of Finnart…
>
> The **Thistle** is a screw steamer of beautiful model, and experienced nautical men who have examined her predict that she will combine the essential elements of speed and steadiness. Her dimensions are as follows: length of keel and fore-rake 190 feet; breadth of beam 25 feet; depth of hold 14 feet; tonnage 600 tons. She will be propelled by a pair of geared beam engines of 150 horsepower, which have been constructed by Messrs A & J Inglis, engineers of this city. Immediately after her launch the **Thistle** was towed up to Glasgow to receive her machinery. The vessel will be employed in the Glasgow, Londonderry and Sligo trade, and will be neatly and substantially fitted up for the carriage of cabin and steerage passengers, and for cattle and goods.
>
> After the disposal of the important business of the launch, a party of about fifty ladies and gentlemen sat down to an elegant repast in an apartment in the works which had been prepared for the occasion - Mr Walter M'Lellan in the chair and Mr John Cameron, croupier…

The *Thistle* unusually had an American-style geared beam engine connected to a single screw. When she entered the Londonderry service, the *Londonderry* was withdrawn and sold for demolition. On 13 December 1859, the *Thistle*, bound up the Clyde, came in collision with the *Henry Bell,* while the latter was leaving Victoria Harbour for Liverpool. The *Thistle* was largely unscathed, but the *Henry Bell* sustained serious damage and had to put back to land her passengers and crew for fear of her sinking.

On 27 December 1859, the *Shamrock* towed the disabled Belfast to Fleetwood paddle steamer *Lyra* into Greenock. The *Lyra* had been in contact with the schooner *Sweetheart*, the latter loaded with coal for Belfast, off the Mull of Galloway. The *Sweetheart* sank rapidly, taking three of her five crew members with her, while the *Lyra* had a severely damaged starboard paddle box. The passengers aboard the *Lyra* were transferred to the passing barque *Patience* and taken to Ardrossan.

In 1859 departures from Glasgow to Londonderry were advertised to leave every Tuesday, Thursday and Saturday, the Thursday sailing calling also at Portrush. Departures were in the afternoon or early evening from Glasgow, calling also at Greenock to connect with the train from Glasgow, which left 2 hours after the ship had left the Broomielaw. The cabin fare, including the steward's fee, was 12/- and the steerage fare was 3/-. The Sligo boat left Glasgow every Saturday afternoon or evening and returned from Sligo on Tuesday morning, normally with a 7 am departure. G & J Burns' daily service to Belfast now offered connecting and competing train services to and from both Londonderry and Dublin.

A new *Myrtle* was completed in 1859 and registered under the ownership of the London & North Western Railway but managed by T Cameron & Company. She was put onto the Poulton-le-Sands to Londonderry service.

New Directory of the City of Londonderry 1839

12 May 1843. Sligo. **Rover** Thursday 18 May, then fortnightly. Thomas Martin & Burns & Co.
7 February 1845. Sligo. **Rover** Friday 14 February. Thomas Martin & Burns & Co.
1 August 1845. Sligo. **Rambler**. Thursday 7 August.
9 January 1846. Port Rush. New steamer **Saint Columb** every Monday. T Martin & Burns & Co.
30 January 1846. Sligo. **Rover**, Friday 30 January.
6 March 1846. Sligo. **Rambler**, Saturday 14 March and Monday 30 March.
12 June 1846. Sligo. **Rambler**.
19 June 1846. Port Rush, calling at Larne once a fortnight. **Saint Columb**. Every Monday, returning every Thursday.
2 July 1847. Sligo. **Rover**, T Martin and Burns & Co. Also the new **Shamrock**.
21 September 1847. Sligo. **Shamrock** (new steam ship).
3 December 1847. Londonderry and Sligo. **Shamrock**, alternate Mondays.
28 January 1848. Sligo. **Shamrock**.
3 March 1848. Sligo. **Rover**, T Martin and Burns & Co.
16 May 1848. Sligo. **Shamrock**, Thursday 25 May and 1 June.
23 May 1848. Sligo. **Shamrock**, Thursday 25 May and 1 June. T Martin and Burns & Co.
1 August 1848. Sligo. New steam ship **Londonderry**, Tuesday 1 August. T Martin and Burns & Co.
8 August 1848. Sligo **Londonderry**, Thursday 17 August and Friday 25 August.
18 August 1848. Sligo. **Londonderry**.
13 October 1848. Liverpool and Port Rush. **Saint Columb**, T Martin and Burns & Co.
19 December 1848. Liverpool and Sligo. **Shamrock**.
9 January 1849. Liverpool Sligo. **Shamrock**
16 January 1849. Liverpool and Port Rush. **Saint Columb**, Thursdays 18 January and 25 January. Leave to call at Larne.
2 March 1849. Liverpool, Port Rush & Sligo. **Shamrock** and **Saint Columb**, this is a combined advertisement but **Shamrock** sails to Sligo, and **Saint Columb** sails to Port-Rush. T Martin and Burns & Co.
6 April 1849. Liverpool to Port Rush/Sligo. **Shamrock** to Sligo; **Rover** to Port Rush (leave to call at Larne).

CHAPTER 4

EXIT MR CAMERON

In 1860 Thomas Cameron was still very much in charge of T Cameron & Company, managing agents of the Glasgow & Londonderry Steam Packet Company. John Cameron, son of Thomas, had become a senior partner. The directors of the Steam Packet Company included George Burns as a former shareholder, Walter MacClellan and Archibald MacConnell. The company was expansive, but had a great respect of risk taking. It maintained three departures a week to Londonderry, with one ship calling at Portrush, a weekly service to Sligo and a fortnightly service between Sligo and Liverpool. These services were maintained by the paddle steamers **Shamrock, Rose** and **Myrtle**, and the screw steamer **Thistle**. The screw steamer **Garland** spent the spring months on the charter market, part of the time to M Langlands & Sons on their Liverpool to Glasgow service. The company also managed the Morecambe screw steamers **Arbutus** and **Myrtle** for the 'Little' North Western Railway which was now worked and managed by the Midland Railway.

The rival McConnell and Laird service between Glasgow and Londonderry was awaiting delivery of the new steamer **Falcon**. It was maintaining its service with the screw steamer **Silloth**, built in 1856 and chartered from the Solway Bay Steam Navigation Company. She sailed from Glasgow on Monday and Friday afternoons and, weather permitting, called at Portrush. In addition, the agency at Greenock was still maintained under the name Alex. Laird & Sons. Alexander A Laird also had the steamer **Irishman** working between Glasgow and Dublin.

The **Falcon** was launched on 6 April 1860 from the yard of Archibald Denny at Dumbarton. She was registered under Alexander Laird's name for the City of Glasgow Steam Packet Company. The former company of that name had previously been absorbed by MacConnell & Laird. On 12 May 1860, during her fourth voyage to Londonderry, she stranded in fog on Patterson's Rock, Sanda, off the Mull of Kintyre, but was refloated the following day. The subsequent inquiry determined that the accident was caused by the compass being affected by additional iron guard railings being installed near the bridge. The **Falcon** quickly returned to service.

During 1861 the Midland Railway withdrew its support of the Morecambe to Londonderry service. Following the Railway's withdrawal, the service was then operated by **Rose** and **Thistle** under the auspices of the Glasgow & Londonderry company.

The collective fleet had quite an eventful start to the decade; Lloyd's List 2 April 1861, reported:

> Belfast 30 March. The **Olive Branch** (brigte.), Russell, of and for Belfast, from Maryport, with coals, has been assisted to the Queen's Quay this morning from off the north side of the Victoria Channel, where she had been run ashore with 5 feet of water in her, after collision with the **Arbutus** (s.s.), of Lancaster, for Morecambe; the water had reached over her deck shortly after she was put ashore.

The **Garland** was in trouble on the Clyde twice in November 1861. On 6 November the lighter **Ann**, of Bowling, was coming down the river in tow, when she was run into and sunk near Garvel Point by the **Garland** coming in from Sligo. Only a portion of the masthead of the lighter was visible above water. One week later, on 13 November, the **Garland**, having called at Greenock, was going up river when she ran into the centre of a tow of vessels going from Bowling to Greenock, behind the tug **Success**. One of the vessels, laden with coal, sank, and another received considerable damage.

Then, as reported in Lloyd's List, 22 October 1862:

> Morecambe 20 October. During the heavy gale last night the **Laurel** (s) broke adrift, and ran into the **Garland** (s.s.), doing the latter considerable injury. The **Laurel** carried away her own head, cutwater, bowsprit, etc., and damaged the stern.

…and in Lloyd's List, 2 June 1863:

> Portrush 30 May. Yesterday, as the **Arbutus** (s.s.) was coming into harbour, she got in collision with the **James Annand** (schr.), and carried forestay away, splitting knight-head, low rail, etc.

The **Thistle** was a notably fast ship capable of maintaining a sea speed of 13 knots. In 1862 she caught the attention of the Confederate agents looking for suitable ships to purchase and modify as blockade runners during the American Civil War. She was bought by George Wigg of Liverpool and work was put in hand to make her ready to cross the Atlantic. It was clearly a profitable deal for the ship owners; a Dispatch from the US Consulate at Liverpool, dated November 7, 1862, read:

There are now nine vessels loading at this port for the rebels, all of which will either attempt to run the blockade or else land their cargoes at Nassau, Bermuda, St Thomas or Havana, to be run into Charleston on the steamers running between Charleston and those islands, viz: **Thistle**, **Nicholas III**, and **Julia Usher** (steamers); ship **Digby**; barks **Severn**, **Queen of the Usk** and **Intrinsic**; also bark **Mary Francis** and ship **Monmouth**. These are all English vessels [sic] and will sail under the British Flag. English steamers in port ready to load: **Peterhoff**, **Gladiator**, **Minnie**, **Bahama**, and **Stanley**.

Thistle did cross the Atlantic but on 8 March 1863 she grounded off Battery Beauregard, while leaving Charleston, and was captured by USS **Canandaiqua** and USS **Flag**. She was salvaged and renamed USS **Cherokee** in May 1863.

The Camerons realised the profitability of building another ship specially for the blockade runner scouts to buy. A fast paddle steamer was ordered from Laurence Hill at Port Glasgow. She was of composite build with a wooden hull on an iron frame, allowing for rapid construction. The ship was launched as another **Thistle** in June 1863, and registered under the ownership of John Cameron and Lewis MacLellan for the Glasgow & Londonderry Steam Packet Company. She was not as fast as her namesake, but could manage a healthy 10 knots. She was rapidly bought by agents in Liverpool who converted her ready to be loaded with military hardware and foodstuffs, for the Confederate Army. She sailed early in 1864 but was captured on 4 June in a position east of Charleston by USS **Fort Jackson** and taken into the US Navy as USS **Dumbarton**. The US Department of the Navy - Naval Historical Center records that:

> **Thistle**, a 636-ton iron side-wheel steamship, was built at Glasgow, Scotland in 1863. She was operated as a blockade runner during the American Civil War, making a successful round-trip voyage between Bermuda and Wilmington, North Carolina in March-May 1864. However, another attempt to run into Wilmington was cut short when she was captured by the USS **Fort Jackson** on 4 June. The US Navy purchased her from the Boston Prize Court in July and, after conversion to a gunboat, placed her in commission as the USS **Dumbarton** in August, 1864. After participating in a fruitless search for the Confederate raider **Tallahassee**, she returned to the Wilmington area where she served until December as an enforcer of the blockade that she had previously challenged. The **Dumbarton** had flagship duty on Virginia's James River during February and March 1865 and was subsequently decommissioned at Washington, DC. Moved to New York in November 1865, she was sold to the Quebec & Gulf Ports Steamship Company in October 1867. Briefly retaining the name **Dumbarton**, she changed to British registry and was given a new rig and engines in the months after she left Navy ownership. During 1868-1870 she operated under the name **City of Quebec**, but was sunk in a collision while in Canadian waters…

A third ship was destined for the Confederate agents, the **Laurel**, launched by A & J Inglis at Pointhouse on 3 September 1863. Again the registered owners were John Cameron and Lewis McClellan, on behalf of the Glasgow & Londonderry Steam Packet Company. The ship took up service at the end of the year running from Glasgow to Sligo and Liverpool to Sligo. When the **Laurel** was leaving the Clyde bound for Sligo on 9 February 1864, she struck the Rothesay steamer **Petrel** carrying the paddle box away. The **Laurel** had her bow damaged in the incident. Then, on 31 August 1864, the **Laurel** was in the Mersey, bound for Sligo, but had to be beached at Egremont after one of her valves burst. The ship was towed off safely the next day and docked. The commercial service of the **Laurel** ended in November 1864 when she was bought by Henry Lafone, Liverpool, for the Confederate Government as another blockade-runner. She was renamed **Confederate States** under the command of Lieutenant John Ramsey. He made just one successful run through the blockade before the civil war ended, and the ship soon found her way back to Liverpool and onto the commercial market.

Just how much money had been put into the bank in Glasgow by the Confederate agents through these three windfall deals is not recorded, but it would certainly have been welcome.

The **Irishman** was damaged in collision off the north western coast of Ireland on 16 September 1863 with the **Barbara Campbell**, of Glasgow. She was arriving from Grenada and bound for Greenock, with sugar and rum. The railway steamer **Duke of Cambridge** came to the assistance of the **Barbara Campbell** and took her in tow until she sank off the South Rock; one man was lost. The **Irishman** was put ashore outside Ardglass Harbour, having partially filled with water. Divers were put to work to salvage the cargo and by 28 September she had been raised and secured in the harbour, just before the weather deteriorated. The vessel was patched up and sent to the Clyde for repairs and, at the same time, lengthened by 17 feet. During this work William Sloan's steamer **Corsair** was chartered to maintain the Dublin sailings.

On 16 May 1864, the **Garland** stranded on Rathlin Island on a voyage from Glasgow to Sligo. A tug was sent to assist her while all passengers were safely landed ashore. It was later reported that she was filling with water but that the holds were largely watertight and the cargo had been removed undamaged. The weather was kind to the ship and she was refloated and brought up to Londonderry on 19 May for repairs.

A new **Myrtle** was delivered by Barclay, Curle & Company early in 1865 and put onto the Sligo and Ballina services. Ballina and Westport, on the west coast of Ireland, had been occasional ports of call for the smaller ships since the early 1850s but for cargo only. The **Myrtle** was a small cargo ship designed specifically for the livestock trade. On arrival at Sligo on 20 March, she reported being struck by a large wave off Fair Head, which started some of the plates, causing her to make a little water which in turn damaged some of the cargo in her after hold. Nevertheless, she was able to sail for Glasgow the following day.

Three months after the delivery of the **Myrtle**, a new **Thistle** was delivered to the Glasgow & Londonderry company. She was a much more substantial passenger and cargo steamer than the **Myrtle**, and was registered in the names of John Cameron & Lewis MacLellan as Trustees. By all accounts she was a handsome steamer with three masts, and she quickly became popular on the Londonderry service.

The **Arbutus** went ashore on a Londonderry to Morecambe voyage on 3 April 1865, between Port Ballintray and the Giant's Causeway. Her master, Captain Jacobson, was able to summon a tug for assistance but, by 6 April, the ship had filled to the beams and attempts to salvage her were abandoned. Efforts were then focused on saving as much of the cargo as was possible. Despite this grim situation the vessel was floated off and the ship was brought to Londonderry on 26 April where she was surveyed to assess the damage.

A new port of call was added to the Glasgow & Londonderry services when the **Myrtle** first arrived at Coleraine from Glasgow on 11 May 1865. This was welcomed by traders in the town who hoped this was the start of a regular service. However, navigation of the shallow bar at the entrance to the River Bann proved so hazardous that the call was dropped in August of the same year. Another new service introduced in 1865 was between Portrush and Liverpool via Larne with occasional calls at Garlieston, Wigtownshire. This service, however, was maintained for the next 35 years, and the company invested in the Portrush Harbour Company in order to safeguard its position there. The company already ran a fortnightly service between Westport and Liverpool via Portrush and Larne, which had commenced in 1864.

Lloyd's List reported on 27 June 1865:

> Dublin 24 June. The **Janet Taylor** (smack), of this port, Flanagan, bound to Penzance, with potatoes, was about to proceed from the Ballina River 16 May, when she was struck amidships and sunk by collision with the **Garland** (s), of Glasgow, which proceeded to Ballina; crew saved.

The *Derry Journal* of Wednesday 20 September 1865, included the lead headline 'The collision in Lough Foyle between the **Garland** and **Falcon**'. The column begins: 'On Saturday evening, about half-past six o'clock, news was received from different sources that a collision of a somewhat important character had taken place in Lough Foyle.' The **Garland**, with a cargo of livestock and about 50 passengers, was heading down the Foyle from Londonderry quay bound for Glasgow. Between Quigley's Point and Whitecastle the **Garland** struck the **Falcon** on the port bow, 'cutting her down almost to the water's edge.' The **Falcon** was heading up the Foyle towards Derry 'crowded with Irish reapers returning from the Scotch harvest.' Two men died from the impact and a further fifteen were drowned. Both ships were repaired and soon back in service.

On 7 December 1866, the **Falcon** was in trouble again when she ran into the smack **Catherine Mary**, of Crarae, Lochfyne, at the Tail of the Bank. The smack sank rapidly. She had been on passage from Furnace, near Inveraray, to Greenock. The **Falcon** received no material damage and afterwards proceeded on her way to Londonderry. Just a month later the **Falcon** herself was lost. On 8 January 1867, the **Falcon** was stranded near Machrihanish, while seeking shelter during a snow storm off the Mull of Kintyre, during a voyage from Glasgow to Londonderry. The ship slid off the rocks and sank about a mile offshore. Only the captain, the second mate Hugh O'Donnell and winchman James Ure survived – 20 other crew members, with 16 steerage passengers and one cabin passenger, also took to the boats but all perished. The wind was south south east Force 9. The *Glasgow Herald* reported the loss on 9 January 1867:

> The **Falcon** should have left Glasgow for Londonderry at midday on Friday, but owing to the fog which prevailed on the river she was detained till the following day when she dropped down the river with a cargo of light goods. Arrived at Greenock she loaded a quantity of sugar and then sailed for Londonderry at half past twelve on Saturday afternoon under the command of Captain Richard Hudson. She passed the lighthouse on the Mull of Cantyre [sic] about six o'clock the same evening. Proceeding on his course from this point the captain intended to take the channel between the island of Rathlin and the mainland of Ireland. By the time, however, that he had, as he reckoned, run the distance, or nearly so, the night had become very thick, with heavy snow and sleet, the wind having at the same time increased to a strong gale blowing from south by east. Under these circumstances, Captain Hudson deemed it expedient to put about in order that he might lie to shelter of the Cantyre land. Following out this plan he continued to keep the vessel's head to the wind, going dead slow, till about 2 o'clock on Sunday morning when she struck the land somewhere between the Mull of Cantyre and Macrihanish Bay, on the west coast of the peninsula.

> On finding himself aground, Captain Hudson ordered his engineer to back the engines, in order to carry the ship clear of the rock or shore with which she was in contact. For a short time she stuck fast, defying the power of the engines to move her, but at length the action of the waves lifted her off. The captain then directed the engineer to

start the engines ahead; his object being, if possible, to run ashore in a small creek or inlet which he thought he observed through the darkness, and in which he considered he might beach his vessel in safety. Unfortunately, when the engineer applied himself to the machinery he found that some water had accumulated in the cylinders, thus preventing them for a time from working. This appears to have been the critical moment, the losing of which proved disastrous. Before the water could be blown out of the cylinders the vessel's head had fallen off and her stern had been turned towards the land. When the engines were at length got into motion the captain ordered them to be backed, with the view of running ashore stern foremost. By this time, however, owing to the damage which had been done to the fore part of the vessel, the bows had sunk considerably, and the stern was consequently raised to such an extent that the screw had no power to back her.

All hope of beaching the steamer being thus at an end, the crew began to think of the boats. These, we understand, were quite sufficient to have taken ashore twice the number of people on board; but as far as we have yet ascertained only one party left the vessel…

It appears that the four boats were swung out but as the first boat was lowered it was swamped and had to be brought up again to be bailed out. A second boat was launched by the second mate and a fireman ready to board the remaining crew and all passengers. Not one person came forward to board the lifeboat, all feeling safer on the ship despite her half sunken state. As the water reached the bridge Captain Hudson was persuaded to jump for the boat but no-one else followed. As the **Falcon** began to go under with its human cargo, the painter of the lifeboat was cut. The three survivors were at sea, drifting for twelve hours before they landed at Islay to report the catastrophe. The Lloyd's agent at Campbeltown had no information of the wreck which had sunk in deep water.

Chartered ships were able to continue the City of Glasgow Steam Packet Company's service to Londonderry. However, the service was formally closed in June that year. As it happened, MacConnell & Laird had a new ship on the stocks at Blackwood & Gordon's yard at Port Glasgow at the time of the loss of the **Falcon**. She was launched on 19 June as **Erin**, a passenger and cargo steamer designed for the Glasgow to Londonderry service but completed for the Glasgow & Londonderry Steam Packet Company as **Rose**. At about the same time the **Jamaica Packet,** built in 1864, was acquired and renamed **Scotia**; her registered owners were initially Alexander A Laird and Glasgow ship owner Martin Orme for MacConnell and Laird. In July 1867 her ownership was also transferred to the Glasgow & Londonderry Steam Packet Company and she was renamed **Laurel**.

So what had happened to cause the closure of a viable service and ostensibly leave it in the hands of the Glasgow & Londonderry company? Events had been precipitated by the death of Thomas Cameron which left son John Cameron as sole owner. John Cameron, it would appear, had little appetite for the business and was in any case in poor health. He was approached by Alexander A Laird (junior) in March 1867 with an offer to buy his majority shares in the Glasgow & Londonderry Steam Packet Company, with the agreement that the **Erin** and the **Scotia** be part of the deal. The two men had never seen eye to eye and it must have been a sad day for Cameron when he agreed to the deal. At the same time it was also agreed by the directors that Laird should assume the agency at Glasgow as the new managing agent, and that for a transitional period the agency be carried out under both the old Cameron title and the new name of Laird.

The goodwill and assets of Alexander Laird's Dublin service were also to be transferred into the Glasgow & Londonderry Steam Packet Company. However, the identity of the Dublin service remained independent and, to all intents and purposes, it was still Alex Laird & Company. It even retained its white funnel and black top inherited from the St George company that once used to serve the Glasgow to Dublin route. The parent Glasgow & Londonderry company had used a funnel colour of red over white over red bands beneath a black top for many years, but the red base was phased out at about this time.

In a further twist to the agreement between Cameron and Laird, it was also agreed that fellow director George Burns step into the breach left by the cessation of MacConnell and Laird and be encouraged to run a twice weekly livestock and passenger sailing from Londonderry into Glasgow. This may seem a strange business ploy today, but it was common at that time to group competing interests, especially in the coasting trades. It would have been seen as safeguarding the service should either company be unable to trade. A longstanding example of this kind of arrangement was the shared Liverpool to Glasgow service operated by Messrs G & J Burns and M Langlands & Sons.

Thus, in a rapid turn of events, Alexander A Laird & Company had taken over the Glasgow & Londonderry Steam Packet Company and closed the MacConnell & Laird interests down. Not only had Alexander A Laird & Company become majority shareholder of the Glasgow & Londonderry company but they had also assumed the agency at Glasgow.

Alexander Laird's experience in the trade and in business management was second to none. He had diverse interests over the years and could see opportunity as it arose. An announcement nine years previously in the *Glasgow Herald*, 13 August 1858, is a typical example of this diversity, showing how Laird had been instrumental in developing trade to the east coast and how he sought to move on, building from one business activity to the next:

The interest of the subscribers, David Smith and Alexander Allan Laird, in the concern carried on as Traders between Greenock, Glasgow, Leith and Kirckaldy [sic] under the firm of The Clyde and Forth Steam-Shipping Company, ceased in terms of the Contract of Co-partnership, on 31 July 1858, from which date they have withdrawn from said concern…

Laird was also highly respected and had developed supportive groups of shippers and merchants, not only in Glasgow and Greenock, but also in Londonderry and Dublin. Laird also had a great respect for George Burns and it was clear from the outset that competition between the Laird interests and the Burns interests was to be avoided, and that each company would support the other in any way it could.

One of the first business deals to be struck by Laird in his new role as managing agent for the Glasgow & Londonderry company was with the Sligo Steam Navigation Company. The two companies worked in friendly negotiated competition on both the Sligo to Glasgow and Sligo to Liverpool routes, with services worked on different days or different weeks for the benefit of the shipper. Thus, in order to save the Sligo company the expense of an agent at Glasgow, and the Laird company an agency at Liverpool, the Sligo company was given sole rights to the Liverpool service and Laird sole rights to the Glasgow service. This was a sensible and amicable agreement made for the benefit of both companies. In addition the **Garland**, which had been the main steamer on the Sligo to Liverpool run, was sold to the Sligo Steam Navigation Company in May 1867 and renamed **Glasgow**. At first she was registered under the ownership of its Trustees, Messrs William Middleton and William Polloxfen, but after a brief period she was registered in the ownership of the Sligo Steam Navigation Company.

'D & G Steam Co.' in a belt logo; Dublin & Glasgow Steam Packet Company demitasse spoon.

CHAPTER 5

ALEXANDER A LAIRD IN CHARGE

There were no immediate changes to the operation of the Glasgow & Londonderry Steam Packet Company following the sale in March 1867 of the majority shareholding by John Cameron to Alexander Allan Laird. The only significant change was the termination of the direct Morecambe to Londonderry service at the end of the year. This service had been operated by *Rose* and *Thistle* since 1861 when the Glasgow & Londonderry company had assumed responsibility for the service after the Midland Railway had withdrawn. The only consequence of the termination of this service was the sale of the *Myrtle* to London owners (she was repurchased by Alexander A Laird four years later).

The *Rose*, *Thistle*, *Shamrock* and *Laurel* maintained the Londonderry, Portrush, Sligo and Ballina services. Departures to Londonderry were on Mondays, Thursdays and Fridays, returning from Londonderry on Tuesdays, Wednesdays and Saturdays. Departures were fixed at 4 pm from Greenock and 6 pm from Londonderry. Glasgow sailings to Portrush were on Wednesdays and Fridays with return sailings to Glasgow on Mondays and Thursdays. Sligo departures were approximately weekly and the saloon fare was 12/6d, as on the Londonderry service, but the steerage fare was more expensive at 5/- rather than 3/- to Londonderry. Separate sailings to Ballina were scheduled roughly every fortnight 'with liberty to call at Sligo'. The *Irishman*, still under the MacConnell and Laird agency in December 1867, was advertised to sail for Dublin every Tuesday, returning to Glasgow every Thursday. The cabin class fare was 10/- and steerage 5/-. The new Burns service between Glasgow and Londonderry generally saw the steamer *Weasel* departing Glasgow on Wednesday and Saturdays. She called at Greenock and a single cabin class passage was offered at 12/6d and steerage at 3/-, the same prices as offered by the Glasgow & Londonderry company under strict Conference protocol.

A passenger and cargo steamer was launched from the yard of Randolph, Elder & Company at Govan on 31 August 1867 and christened *Vine*. She never sailed for the Glasgow & Londonderry company, as she was completed for Donald R Macgregor and others, of Edinburgh, and employed in the Baltic trade.

The paddle steamer *Bridgewater* was bought in February 1868 from the Liverpool & Dublin Steam Navigation Company and renamed *Garland*. She was placed on Morecambe duties when the Morecambe to Londonderry service was resumed in September 1868. She had been ordered by G & J Burns as *Roe* and was intended to run on that company's Glasgow to Belfast service, but was completed as the blockade runner *City of Petersburg*, returning to Glasgow in 1865.

On 12 September 1868 the *Rose* and the schooner *Mary Jane*, on passage from Bowling to Tobermory with a cargo of coal, were in collision off the Garmoyle light. The *Mary Jane* sank quickly, although her crew was taken safely aboard the *Rose*, which was barely scratched.

On 26 November 1869 the *Rose*, sailing for Londonderry, collided with the Anchor liner *Cambria*, arriving in the Clyde from New York. The story aptly summarises the hazards of travel by sea in this period; the *Glasgow Herald*, 27 November 1869, reported:

> About 1 o'clock yesterday morning, a collision of a very serious nature occurred in the Channel, nearly abreast of the island of Pladda, which although happily unattended with the loss of life or injury to person, resulted in such damage being done to the Londonderry and Glasgow screw steamer *Rose*, as to necessitate her being run ashore to the eastward of the above named island, in order to prevent her from foundering.
>
> From the details which we have been enabled to gather, it appears that the SS *Rose*, 278 tons, under command of Captain Galbraith, left Greenock Quay about 9 o'clock on Thursday night for Londonderry, having on board a general cargo and about forty passengers, only five of whom were in the cabin. Several of those on board, it may be remarked, intended to disembark at Moville, and go aboard the *Nova Scotian* bound for Canada. The weather was moderate, with a slight breeze from the south west. There was an occasional shower of rain, but the atmosphere was clear, the moon shining softly overhead. Everything went well till Pladda Island, to the westward of Arran, was reached, when unhappily a collision occurred betwixt her and the inward bound American screw steamer *Cambria*, Captain Craig, which resulted in serious damage to the *Rose*, while the other vessel escaped with comparatively little injury.
>
> The *Rose*, it is stated, was steering her usual course, keeping somewhat close to the Arran land, while the *Cambria* had all the rest of the Channel clear, the course for a vessel coming up Channel being in no way obstructed. Those on board the *Rose* observed the coming of the large steamer about half an hour before the collision took place, and it was likewise seen that the approaching vessel was under a press of sail, and was keeping close upon Pladda, the danger whistle was sounded to warn those on board of the neighbourhood of an outward bound steamer. All the

lights were burning at the time. Observing that the **Cambria** (which the inward bound steamer afterwards proved to be) was still heading inshore, the **Rose** was steered as close to the island as was considered prudent, but the **Cambria** continued to starboard her helm, and when nearly opposite the lighthouse which is erected on that island, and a few hundred yards to the south of it, the **Cambria** bore down upon the **Rose**, and struck her a few feet aft [of the bow] on the port side, cutting her down below the water line. The shock was a very violent one, and was heard at some distance.

The alarm which prevailed amongst the crew and passengers was of course very great, and the wildest consternation was for some time exhibited, especially amongst the females. On board the **Cambria**, which had a complement of 109 passengers, there was also a feeling of great alarm. Most of them turned out of their berths lightly clad, in order to ascertain the cause of the fearful shock the ship had just experienced. As we have already stated, however, no lives were lost and as far as is known, no person was injured. Five of the crew and three of the male passengers on board the **Rose** succeeded in scrambling over the side of the **Cambria** ere the vessels separated. The engines of the **Cambria** were at once stopped, and the vessel having been hove to, a boat was quickly lowered and dispatched to the **Rose**, in order, if possible to render assistance. Fortunately, two tug-steamers, which were lying to eastward of Pladda, upon the collision taking place, at once bore down upon the vessels. The passengers on board the **Rose** were given over to one of the tugs, which afterwards proceeded to Lamlash, where they were transferred to the steamer **Lady Mary**, Captain Brown, and carried to Ardrossan. Here they took a train for Glasgow and arrived in the city at four o'clock in the afternoon. The other tug, we may state, went to the **Cambria** with a view to rendering assistance, which, however, was happily not needed.

Immediately on the occurrence of the collision the forward compartments of the **Rose** filled, and she grounded at the bow. She was afterwards, however, brought round at the stern by means of ropes attached to one of the tugs, and being floated off at the bow, she was beached stern foremost in a sandy and sheltered bay at the east side of the island. Her bow sank in the sea down to the bulwarks. The **Cambria**, after it had been ascertained that no assistance could be given to the disabled ship, was taken to the Tail of the Bank, which she reached yesterday morning. She afterwards proceeded to Glasgow, where she arrived shortly after six o'clock, and was berthed near Lancefield Quay. She has lost her figurehead, and one or two plates have been started. On information of the occurrence being received, Mr Weild, Underwriter's Surveyor, proceeded to Pladda on board the steam tug **Flying Spray,** in order to arrange for having the **Rose** brought to some port for repairs. Steam pumps were to follow him last night.

Three days after the incident the **Rose** had been refloated and brought onto the bank at Lamlash Quay by the tugs **Stork** and **Gem**. *Lloyd's List* reported on 30 November 1869:

Lamlash. The whole of the cargo from the **Rose** (s), from Glasgow to Londonderry, which was brought here after being ashore at Pladda, after collision, has been discharged, and has received no damage. The vessel is cut down to within four feet of her keel, about twelve feet from the bow.

The *Glasgow Herald* stated on 6 December 1869:

On Saturday afternoon the Glasgow & Londonderry screw steamer **Rose**… arrived at Greenock under steam, and after landing Mr Weild, Underwriter's Surveyor, proceeded direct to Glasgow, where she will be surveyed and placed in dock. It will be remembered that the **Rose** received such severe damage on the morning of Friday 26th ult., that it was found necessary to run her ashore on Pladda, to prevent her from foundering…

And the **Cambria**? She had commenced her maiden voyage from Glasgow via Moville to New York only on 8 May 1869. On 19 October 1870, under the command of Captain George Carnahan, on only her twelfth crossing of the Atlantic, she was wrecked on the Donegal coast with large loss of life.

The **Laurel** was in trouble in navigating the Clyde shortly after the **Rose** incident. *Lloyd's List* 30 December 1869, reported:

Greenock 29 December. The **Laurel** (s.s.), trading between the Clyde and Portrush, ran on the dyke above Dumbarton Castle, whilst coming up the river, during a fog on the evening of 27 December, and remains, an attempt to get her off at high water having failed; a tug and a lighter were yesterday sent hence to take out part of her cargo.

As with the **Rose**, she was later taken off the bank safely and repaired, ready to resume Portrush sailings.

The **Irishman** was relieved on the Dublin service from Glasgow in December 1869 by William Sloan's steamer **Oscar**.

In 1870 a new service between Fleetwood and Londonderry was started by the Lancashire & Yorkshire and London & North Western railways. Two paddle steamers were employed on the route, the venerable **Royal Consort,** built for Kemp & Company in 1844, and the former blockade runner **Old Dominion**, which was renamed **Sheffield**.

The *Rose* suffered two minor groundings in quick succession. On 2 August 1870 she ran ashore at Fort Ballantrae on passage to Londonderry, after colliding in dense fog with the schooner *Sarah* of Belfast, which the *Rose* then took in tow. The *Sarah* had cutwater, bowsprit and headgear carried away, and the two ships went aground together, but came off on the flood and proceeded to Portrush. On 29 April 1871 the *Rose* was again aground, this time on the dyke opposite Dumbarton, coming in from Portrush. She was towed off the Bank on 1 May, and taken to Glasgow to be put on the slip for examination; only minor damage was found.

Glasgow and the upper Clyde were enveloped in a thick fog on 24 January 1871, in which the *Irishman* got into trouble, as reported in the *Glasgow Herald* 25 January 1871:

> The city was enveloped in a dense fog yesterday, and in places of business it was found necessary to burn gas the whole day…An almost entire stoppage was put to the upward and downward traffic on the river, only one vessel managing to reach the Broomielaw. A great many ships are lying below Renfrew unable to make any progress. The Anchor liner *Anglia* left the harbour early to proceed down to Steele's Dock at Greenock… but up to a late hour she had not passed Bowling. The only accident we have heard of occurred in the afternoon. The ss *Irishman*, while proceeding down to Greenock, whence she intended sailing for Dublin, came into collision, when near Bowling, with the ss *Kintyre*, which sailed from Campbeltown on Monday, and was due here today. Both vessels sustained damage, the *Irishman* so much so that it was found necessary to beach her.

The *Irishman* was put on Blantyre Bank, where she lay on the bottom, but clear of the navigable channel. The bow of the *Kintyre* was badly damaged but she was able to proceed. Four days later the *Irishman* was successfully refloated and towed up to the Broomielaw where her cargo was unloaded prior to repairs being made to the ship.

The *Myrtle* which had been sold out of the fleet in 1867, was bought by Alexander A Laird from her London owners in 1871, reinstated in the company livery and put onto the Portrush service from the Clyde.

Lloyd's List 15 June 1872, reported:

> Portrush 12 June. The *James Annand*, of Coleraine, hence for Maryport, in beating out of the bay yesterday, missed stays and went ashore outside this harbour, but was got off and proceeded, apparently without damage. The *Myrtle* (s), from Glasgow to this port, in rendering assistance to the *James Annand*, fouled her screw and, in coming for the harbour, took the ground astern, but was hauled off; she afterwards proceeded to Glasgow.

The *Garland* arrived in the Clyde from Londonderry on the night of 25 September 1872, twelve hours late, having encountered a most violent gale from the north north east. Her lifeboat was lost, several stanchions were carried away, and about 150 head of cattle, so it is reported, had to be thrown overboard.

Charter work brought in important income and repeat charter work was even more welcome. Fraser MacHaffie described one such example, with the Larne and Stranraer Steamboat Company, in his book *The Short Sea Route*:

> During the three weeks 6 to 27 December [1872], the [Stranraer-Larne] service was maintained by the paddle steamer *Garland*, of the Glasgow & Londonderry Steam Packet Company… Captain Campbell brought *Garland* from Glasgow and then took *Princess Louise* off to Belfast, reversing the operation when the work on the '*Louise* was completed. . . . . The following year, in *Garland's* second spell on the Larne service, word got round of a free trip to Glasgow, and on Saturday, 20 December, just as *Garland* was about to leave Stranraer, about three hundred piled on board. Captain Campbell put about two hundred of them ashore but allowed the balance to sail up to Glasgow on *Garland*, returning on the Sunday with *Princess Louise*. On the Tuesday before her departure from Stranraer, Captain Campbell had involved *Garland* in an attempt to rescue the crew of a 300 ton brigantine, *John Slater*, of Barrow. When the gale was at its height, a lifeboat was launched from *Garland* for the purpose of securing a hawser to the disabled craft, but the attempt was unsuccessful. The brigantine was lost but the crew of three were safely landed at Stranraer by *Garland*.

> On 7 December 1874, the chartered vessel *Garland* [again on Stranraer-Belfast] could not sail because of wheel trouble, and we read in the cash book that each of the six passengers inconvenienced received a shilling. When *Princess Louise* failed again in February, *Garland* was brought back to cover the sailing.

During most of 1874 the *Garland* had been laid-up in the Gareloch and she was sold following the December charter work.

A new cargo steamer, the *Fern*, was acquired in 1873. She had been built two years previously as the *Gala* for coastal duties on the East Coast. She was equipped with a compound engine and had a whaleback shape designed to shed water. The *Fern* was initially mainly on the twice weekly Glasgow and Portrush service. On 18 September 1873 the *Fern*, at the start of a voyage to Westport, collided with one of the Clyde Authority dredgers. The *Fern* was badly damaged and put back to Glasgow to discharge ready for repairs. The dredger was beached for fear of sinking.

Fern (1871) started life as ***Gala*** and was bought by the Glasgow and Londonderry Steam Packet Company in 1873 and used mainly for the Portrush service.

[Mike Walker collection]

Alexander Laird, managing Glasgow agent and majority shareholder in the Glasgow & Londonderry company, was replaced by Alexander A Laird & Company (Ship Owners) with effect from 13 July 1873. Laird's partners were A R Brand, George Turnbull and William MacConnell.

On 12 November 1873 the ***Rose***, bound for Londonderry, grounded near Dumbarton. Next day, lighters were sent from Greenock to remove part of her cargo and she floated off that evening. On the night of 11 December 1873, the Londonderry and Portrush steamers ***Rose*** and ***Myrtle***, proceeding down river to Greenock, got on the bank during a dense fog; a portion of the cargo of the ***Rose*** was discharged into lighters in order to get her off. About a dozen other vessels also touched the dyke during the fog.

On 1 January 1874 the ***Laurel*** left Glasgow for Sligo and Ballina with general cargo. In poor weather she put into Arran roads for shelter but, having parted both cables, was intentionally beached in Rutland harbour to prevent damage. She was able to get afloat when the storm abated.

The ***Thistle*** arrived at Greenock from Londonderry on 23 October 1874 to report that she had experienced 'fearful' weather. She had on board a full cargo of livestock, but ten cattle and 40 sheep were killed during the storm. On 20 July 1875 the ***Thistle***, coming down the river to Greenock and Londonderry under Captain Chenoweth, collided with the barque ***Champion***, which vessel was lying at the East India Harbour entrance. Part of the bridge of the ***Thistle*** was crushed and part of her rail broken; the ***Champion*** sustained no damage. A little later in the year, on 13 November, the ***Thistle*** became disabled off Portrush and had to be towed to Londonderry by the ***Fern*** under Captain Cameron. Then, on 25 November, the ***Thistle***, sailing for Londonderry, was in collision in the River Clyde with the Glasgow registered steamship ***Albion***, and sustained considerable damage to her port side: her rail, boat and stanchions being carried away. The ***Albion*** had her bow and upper works damaged. At the end of the year, the ***Thistle*** ran aground in Ross Bay in the Foyle on 21 December. She had only just left Londonderry bound for Glasgow and was able to resume her journey undamaged when she floated off.

In late October 1874 the ***Laurel***, under Captain Campbell, lost her foremast in a gale. Then, on 13 December, the ***Laurel***, while on passage from Londonderry to Morecambe, collided off Walney Island with the schooner ***Countess of Selkirk***, of Kirkcudbright. The schooner had her cathead and some bulwarks carried away, and was towed in to Piel by the ***Laurel***, which sustained only minor damage.

Two new ships were commissioned in July 1875. The first to be launched, on 17 May, was the small steamer **Holly** which was designed for the Morecambe service. She was followed into the water four days later by another **Arbutus**, which was intended for the Sligo service. The *Glasgow Herald*, 22 May 1875, reported:

> Yesterday, Messrs. A & J Inglis launched from their shipbuilding yard, Pointhouse, a very handsomely-modelled screw steamer for the Londonderry Steam Packet Company [sic], the principal dimensions being: length of keel and forerake 210 feet; breadth (moulded) 28 feet; depth 15 feet 1 inch. On leaving the ways the vessel was named the **Arbutus** by Miss MacLellan of Blairvadick. Among those who witnessed the launch were Mr Walter MacLellan, chairman of the company, Mr Laird, managing director, Mr M'Nell agent Londonderry, Mr Brand, Mr MacConnell, and others. The **Arbutus** was towed to the steam crane at Finnieston to receive her machinery, which consists of a pair of compound engines of 180 HP nominal, supplied by the builders.

Holly (1875) spent much of her career working the Glasgow to Dublin cargo service.

[Mike Walker collection]

A near sister of the **Arbutus**, the **Iris**, was launched a year later from the same yard, A & J Inglis at Pointhouse. She was christened by Miss Brand of Glasgow. All three ships had the now ubiquitous compound engine connected to a single screw. These new arrivals allowed the **Myrtle** and **Laurel** to be sold out of the fleet in 1877.

The **Thistle** ran into the schooner **Broughty Castle** in Lough Foyle in the morning of 22 March 1876, having left Londonderry for Morecambe. The **Broughty Castle** and her cargo of coal sank immediately, but all her crew were rescued. On 16 August 1876 the **Thistle** ran into the schooner **Elizabeth**, of Belfast, which was anchored in the stream at Londonderry. The schooner was badly damaged; she lost her cathead and the bower chains were broken. The **Thistle** collided with the brigantine **G A Coonan** off Pladda on 16 March 1877. The sailing ship lost her foremast and later anchored at Lamlash. Then on Christmas Day morning the same year, and again off Pladda, the **Rose** ran down and sank the sailing ship **Petrel**, of Belfast. The crew were saved.

The **Irishman** ran aground on Burial Island, County Down on 13 September 1877 on passage from Dublin to Glasgow. *Lloyd's List*, 15 September 1877, reported:

> Belfast 14 September, 5 pm. **Irishman**, 100 yards from Birr Island, heading west, list to port; exposed to winds and sea from WSW to N; 10 feet water amidships, 17 aft; two after compartments tight, others tide rises and falls; weather favourable at present and we are arranging to commence discharge; pumps and tug coming from Glasgow… part cargo might be saved; crew and passengers saved.

By 17 September the weather had held sufficiently for some of the cargo to be removed to lighten ship. Steam pumps were installed in the evening. On 30 September the **Irishman** was floated off the rocks and arrived at Belfast on 5 October after lying in shallow water to await a berth.

Lloyd's List, 12 October 1877, reported:

> Belfast 10 October. The damage sustained by the *Irishman* (s) was as follows: Three plates broken on starboard bilge, 20 feet by 6 feet; port bottom forced up and twelve frames broken. She went into dock and received temporary repairs, and is now out again awaiting her dock turn for thorough repairs; but, as there is likely to be considerable delay, it is intended to employ her in carrying coal during the interval, being sufficiently seaworthy for that purpose.

Once repaired, the *Irishman* was sold to owners in Belfast. The Dublin service was now taken by the *Thistle*, which had been given new compound engines to enable her to reduce the voyage time from 21 hours to about 19 hours. However, before she took up service to Dublin she lost her propeller on 15 October 1877, on a voyage to Westport, but was able to get to Innislyre harbour under sail. She was towed into Westport the next day, after grounding at the entrance to the harbour, no damage was sustained.

For a time in 1878 the Dublin service was taken by *Holly*, which had her funnel painted in the white and black colours of the old St George Steam Packet Company. However, when *Thistle* returned to Dublin later in the year, the City of Glasgow Steam Packet Company brand was dropped in favour of Alex A Laird & Company. Initially the passenger fares of 12/- cabin and 4/- steerage undercut those of the rival Dublin & Glasgow Sailing and Steam Packet Company, which was forced to reduce its prices. Within the year, an agreement was made that an integrated service should be operated. The Glasgow & Londonderry company was to maintain a two ship service sailing from Dublin on Tuesday, Thursday and Saturday with the Dublin company sailing on Monday, Wednesday and Friday.

The *Arbutus* was lost on 18 April 1878 on passage from Londonderry to Glasgow. She hit the rocks in thick fog at Southend on the tip of the Mull of Kintyre.

The *Shipwreck Index of the British Isles* states:

> *Arbutus*. Mull of Kintyre, Southend, near. 55.18N 05.40W. Voyage Londonderry-Glasgow. Carrying a general cargo of farm produce, 209 assorted cattle and some 60 passengers, the passage from Londonderry went well until they ran into thick fog off Northern Ireland. Speed was reduced as they crossed the North Channel, with regular use being made of the signal gun and the foghorn. Just before midnight, with most of the passengers asleep in their cabins, the ship ran ashore north of the Mull of Kintyre lighthouse. There was a degree of panic among the passengers who, in the dark, thought the vessel was about to sink, but were assured by Captain Aitken that they were safe. Four lifeboats were lowered and filled with passengers and crew, two boats rowing directly ashore, the other two going along the coast to the town of Southend. During the night as the tide rose, the ship slid off the rocks into deep water.

On 18 June 1878 the new passenger and cargo steamer *Vine* was launched from the yard of D & W Henderson & Company at Partick. She was designed for the Sligo route from Glasgow. Two near identical sisters were also delivered that year, one the *Cedar*, from D & W Henderson and the other the *Azalea* from A & J Inglis. The *Azalea* was launched on 14 August and her sister followed on 27 November. The *Glasgow Herald*, 15 August 1878, announced:

> Yesterday Messrs. A & J Inglis launched from their shipbuilding yard a beautifully modelled screw steamer for the Glasgow & Londonderry Steam Packet Company, the principal dimensions being 218 ft. by 30 ft. by 15 ft. On leaving the ways the vessel was named *Azalea* by Mrs W M Hyndman of Port Saunders, Victoria. The *Azalea's* machinery, which consists of a pair of compound engines of 250 horsepower, will be supplied and fitted by the builders.

The *Azalea* (1878) at sea, apparently having just stoked up the boilers.

The *Cedar* was christened at her launch by Miss MacLellan of India Street, Glasgow, a daughter of one of the directors. The *Cedar* and *Azalea* each had two compound inverted direct acting engines with cylinders of 36 inches and 68 inches with a 42 inch stroke. They had accommodation fit for 70 saloon passengers. The *Azalea* effectively replaced the hapless *Arbutus*, although *Azalea* started her career working between Glasgow and Dublin. The *Cedar* was principally employed on the Londonderry to Glasgow route, although she also ran to Morecambe.

The *Cedar* (1878) arriving at Portrush.

The last of the paddle steamers, the old *Shamrock*, had served the company since she was commissioned in the summer of 1847. She was finally withdrawn and sold in 1879. The Belfast newspapers described how in 1862 the paddler had put into that port on passage from Sligo to Liverpool in order to 'thaw her passengers out' in the local workhouse, having lost two young children overboard in very rough conditions. Paddle steamer or not, she was clearly an excellent sea boat to have survived the ravages of the north and west coasts of Ireland in winter conditions for so many years. At least, by the late 1870s, conditions had improved for both cabin passengers and more particularly for steerage passengers, but conditions in the latter were spartan and by no means comfortable.

The *Azalea* was joined on the Dublin service in 1879 by a new *Shamrock*. She had 80 saloon passenger berths, the largest complement for any steamer owned by the company to date. This *Shamrock* had compound engines with cylinders of 31½ inches and 70 inches in diameter with a stroke of 45 inches generating 250 horsepower. Similar in many ways to the *Azalea* and *Cedar*, the *Shamrock* was a slightly bigger ship with better appointed saloon accommodation designed to entice passengers away from the competing Dublin & Glasgow Sailing and Steam Packet Company, of Dublin, despite the two companies running a managed schedule with six daily sailings per week. The three new ships were state-of-the-art and represented a significant investment on the part of Mr Laird.

Accidents continued to happen. The *Thistle* left Londonderry on 29 May 1879 for Morecambe, with 150 tons of cargo. Next day she stranded on the sand at the south end of Walney Island in thick fog. She came off again later in the same day without any assistance or damage. On 11 August the *Thistle* left Londonderry at 14.00, with a general cargo for Morecambe. At 19.00, while a dense fog prevailed, she struck a rock off Torr Point, damaging her bow plating and flooding her fore compartment. She was able to put back to Londonderry for repairs. On 15 September 1880 the *Thistle* was on a voyage from Sligo to Glasgow when she encountered a north easterly gale off Tory Island. All the livestock on deck were lost and a number drowned on the main deck.

The *Rose* suffered from a broken stern post in October 1881. On 11 January 1882 the *Rose* lost two blades of her propeller while going down the River Clyde on passage for Greenock and Londonderry. The *Fern* was ashore in the River Moy on leaving Ballina in windy conditions on 11 December 1881. After lightening ship, she was floated off two days later.

The new Dublin service required two ships and *Shamrock* ran alongside either the *Azalea* or *Cedar* until the early 1880s, when any of the larger ships on the Londonderry route were used on the service. Departures from Dublin and Glasgow were on Thursdays and Saturdays and in the summer an additional sailing departing on Tuesdays was added. The cabin fare was 15/- and steerage was 6/-, with slight reductions for return fares.

An advertisement for sailings from Greenock (train from Glasgow) showed the *Fern*, *Iris*, *Vine*, *Holly* and *Rose* sailing at 4 pm for Londonderry on Mondays, Tuesdays, Thursdays and Fridays with a single cabin fare of 12/6d and a steerage fare of 6/-, return tickets were interchangeable with the Burns sailings. Sailings for Portrush were on Mondays and Thursdays at 3 pm with a single cabin fare of 10/- and steerage of 3/6d, and Sligo every Tuesday at 2 pm, continuing to Westport and Ballina, all fares being the same as on the Londonderry route.

The 1870s had seen the Glasgow & Londonderry company go from strength to strength, expanding its routes and sailing frequencies and upgrading its ships. Casualties remained high but considering the narrow approaches to some of the regular ports and the difficulties caused by both rough weather and fog, casualties of some sort or another were inevitable. It should also be remembered that aids to coastal navigation at that time were at best rudimentary, and that the master's personal knowledge of currents, tides and sea conditions at specific locations was critical to the safe navigation of his ship.

CHAPTER 6

REBRANDING AND 'THE LAIRD'S LINE'

A new passenger cargo steamer, the **Brier**, was commissioned in December 1882. She offered cabin accommodation for 60 and was placed on the Londonderry service to run alongside the **Iris**. Like the **Azalea, Cedar** and **Shamrock**, which mainly continued to serve Dublin, the **Brier** was iron-hulled and equipped with a pair of compound engines. She allowed the **Rose**, formerly **Erin**, to be sold to the Ayr Steam Shipping Company who, six years later, gave her the new name **Ailsa** (Chapter 8).

The *Brier* (1882) had accommodation for 80 passengers when she was commissioned for the Glasgow to Londonderry service.

[Mike Walker collection]

The planned launch of a new steamer ordered in October of the previous year was 3 July 1883. The steamer was to be named **Daphne**. The contract price delivered was £18,730, the cheapest of five tenders from five of the best builders. She was specially built for the conveyance of passengers and cattle, had a passenger certificate, and was fitted with steam steering gear, and with all the newest appliances'. The launch was supervised by Mr John Stephen, of Alexander Stephen & Sons, Linthouse, shipbuilders. Mr A R Brand and Mr Turnbull represented the owners. The *Glasgow Herald*, 4 July 1883, described what happened next:

Henry O'Farrell, a riveter says: I was on board the vessel holding the launching flag. I heard the signal given to knock away the supports, and immediately the vessel moved off. I noticed that she was going very smartly, but paid no particular attention to this until, in taking the water, the vessel heeled over to the port side. I threw the launching flag away, and thinking that perhaps the current had caused the vessel to capsize, I ran to the wheel with the Pilot, Wm. Francis, a rigger, and another man whose name I don't know. The Pilot and Francis have both been saved, but I have not since seen the other man. The three of us worked hard at the wheel, so as to counteract what we thought was the force of the current. We turned it round several times, but it did not do any good. The vessel was every moment going deeper down in the water, and getting alarmed for my safety, I jumped off the stern into the water. I can swim, but not very well, I managed to keep my head above water, and shortly afterwards I was pulled aboard one of the tugs which was assisting at the launch. I can't say how many persons were aboard at the time of the accident. There were a good many, the decks being quite crowded.

Alex Cramond says: I am a foreman joiner in the employment of the firm. I was on board the vessel waiting in case my services should be required for anything. I was on the forecastle head. On hearing the supports being knocked away, I paid attention to the manner in which the ship left the ways. She seemed to go all right, but I noticed that she moved very quickly, more quickly than is usually the case. Almost directly the heel touched the water, I saw that the vessel was going to capsize, and I called out to a number of men standing near me 'Look out boys, this is an awful business, we will all be drowned'. The next moment I was struggling in the water. I cannot swim at all, but I came to the surface, and with little difficulty succeeded in getting hold of a way block which was floating near. From this I caught hold of one of the davits of the steamer, and then onto the side of her. She sank gradually, and I remained standing on her until the water was about up to my waist. I was taken off by a tug. I had charge of about twenty men on board, and since the accident I have only seen three of them.

The tragic launch of the **Daphne** (1883); she was later completed as **Rose**.

[From a contemporary engraving]

The specifications for the ship had been drawn up by the Marine Superintendent of the Glasgow, Dublin & Londonderry Steam Packet Company, for service between Glasgow and Londonderry. Alexander Stephen & Sons were responsible for the detailed design and tests on the stability of the hull. As the yard did not have a fitting out quay it was proposed to launch the ship with her engine in place, and with the boiler room left open to the sky, so that the boilers could be installed from a quayside crane up river. The plan was to have her ready for the Glasgow Fair Holiday, so time was all important for the completion of the new steamer. It was for this reason that **Daphne** was launched with nearly 200 artisans on board, all working hard to complete the fitting out by the deadline.

Just before noon the ship was launched quietly into the Clyde and, as planned, the drag chains restrained her onward progress across the channel. The starboard chain worked well but the port chain slipped so that the hull yawed across the channel. Only a small crowd had assembled to watch what, to all intents and purposes, was a routine event on the upper Clyde. Of those present, many had relatives and friends aboard **Daphne**. Within a few seconds of coming to rest, the ship heeled to port, juddered and steadied slightly before she heeled over again onto her side and capsized, trapping many of the workmen below deck. The roll was exacerbated by loose gear, which slid to port and the open deck above the boiler room allowed it to flood in an instant. Rescuers were on hand almost immediately, but there was little to be done other than pulling bodies out of the water; there were only about 70 survivors, while the death toll was later counted to be as many as 124. Many of the apprentices aboard that day were no more than boys.

The ship lay such that the channel was open to a width of about 100 feet. Nevertheless, on the day after the tragic launching, the Allan Line's **Scandinavian**, in coming up the river and attempting to pass between the sunken steamer and the north bank, took the ground and stopped. She was backed off after 15 or 20 minutes steaming and worked into the middle of the stream.

The **Daphne** was righted a few days later on 23 July and taken to Salterscroft Graving Dock at Govan, where the hulk was subjected to exhaustive tests. Alexander Stephen's naval architects were commended for the assistance they gave the subsequent Inquiry; the cause was found to be a lack of initial stability combined with the weight of loose gear and personnel which slid to port, stopping the vessel from righting herself. The hull was later completed as **Rose** for Alex. A Laird & Company and placed on the Londonderry service from Glasgow.

On 2 September 1883 the steamer **Iris** went ashore on Innistrahull island off Innishowen Head at 3 am, on a voyage from Glasgow to Sligo. One man, the cabin steward, was drowned while trying to get ashore. The after deck was completely submerged. On 3 September the *Glasgow Herald* reported:

Details verifying the total loss of the steamship *Iris*, of the Glasgow & Londonderry Steam Packet Company, plying between Glasgow and Sligo, show that the *Iris* was having a good passage and going well at the time of the catastrophe – 3 o'clock in the morning – when suddenly there was a tremendous shock and all on board hurriedly came up on deck… A seaman called Wm. Bruen, of Rosse's Point, Sligo, was the first to ascertain what had happened. The steamship had struck stem on to a rock rising some 50 feet out of the sea. The vessel was then in imminent danger of foundering in 21 fathoms of water. Bruen got onto the rock… got a ladder and made it fast to the rock, and by this means the passengers scrambled out of the sinking ship and clambered ashore as best they could in the storm to the crown of the rock. One death occurred, the steward, John Sharkey, a native of Glasgow, was soon amongst the passengers, but he fell, and was stunned and drowned… As soon as daylight made the disaster known assistance came from the lighthouse and the mainland. The passengers were taken off by tugs. The *Iris* is a total wreck.

News of the disaster was received in Glasgow by the owners, Messrs. Alex. A Laird & Co., about 7 o'clock on Sunday evening. Mr Brand, the managing owner, at once telegraphed to Moville, giving orders for tugs to be sent out to the assistance of *Iris*. The Londonderry agent was also instructed to make preparations for receiving the passengers, and forwarding them to their destinations. There were altogether 20 passengers on board. The ss *Fern*, belonging to the same company, bound for Glasgow, called at Innistrahull in the course of Sunday, and communicated with the passengers, the captain offering to take on board any of them who desired to get to Glasgow. Two took advantage of the opportunity, and were landed at Glasgow yesterday morning, the others preferring to go on to Derry, where they were landed by the tugs yesterday, and forwarded to Sligo by the company's agent…

The *Shipwreck Index of the British Isles* includes an entry which is slightly different from the newspaper report:

2 September 1883. *Iris*. Co. Donegal, Innistrahull, 1m E of the lighthouse. 52.25.50N 07.13W. Voyage Glasgow and Greenock – Sligo. Stranded and lost on the eastern part of the island in wind condition ENE force 6. The agent for Alexander Laird and Company, who operated a number of Irish Sea passenger routes, was attending the evening service in his Londonderry church when a messenger handed him a telegram, and he immediately left to arrange for two tugs to go to the assistance of the s.s. *Iris*. Later on 3 September the tugs *Admiral* and *Triumph* arrived at Derry with some 30 passengers and most of the crew. They told how they had struck a rock separated from Tory Island by a chasm, but three members of the crew had managed to climb the rocks and alert the lighthouse keeper which was only some 400 yards away. The keepers took with them a ladder which was laid across the gap, and with a rope tied round their middle each member of the crew and passengers crawled to safety. One member of the crew, steward John Sharkey, lost his life when he fell off the ladder into the sea.

Salvage operations continued into October and some equipment and a quantity of bacon were landed before the ship and its remaining cargo were abandoned.

On 26 December 1883 the *Rose* (formerly the unfortunate *Daphne*) struck an anchor lying on the bottom in Portrush Harbour and took on a large amount of water. The anchor was removed and the hole patched with box and cement so that the ship could be pumped out; she eventually sailed for Ayr for repairs on 4 January 1884. On the morning of 2 March 1884, the *Rose* went aground on Farland Point near Millport on the Isle of Cumbrae in thick fog. She was inbound to Greenock and Glasgow from Londonderry. Tugs failed to get the ship off her rocky perch and her stern sank deep into sand off the rocks. On 27 March, the *Rose* was floated clear and beached in Kames Bay ready to be pumped out and temporarily repaired before being taken to Glasgow to be put on the slip. Once the ship had been fully repaired, Laird decided he had had enough of her and she was sold to owners in Prestwick. The subsequent Board of Trade Inquiry reported:

She left Port Rush at about 8.30 p.m. of the 29th February, with a crew of 22 hands all told, 6 passengers, and 130 tons of cargo, bound to Glasgow; and at about 11.15 the same evening was off the Mull of Cantyre [sic]. After rounding Paterson Rock, off Sanda Island, which she did at about 12.15, she was put by the master on an ENE course to make Pladda Light; and he then went below, leaving the deck in charge of the second officer. At about 12.45 the chief mate came on deck and took charge, and in due course made Pladda Light bearing a ¼ of a point on the port bow, upon which the helm was by his orders slightly ported to give it a wider berth. After passing it the vessel's course was altered first to NE by E, then to NE, and then to NE by N, until they were abreast of Holy Island Light, when a N by E ¾ E course was steered to make Cumbrae Light. We are told that, when the master went below, the weather was fine; but on nearing the Cumbrae it became thick with drizzling rain and sleet, and thereupon the mate called the master, and on his coming on deck he continued the vessel on a N by E ¾ E course, until they sighted Cumbrae Light bearing a little on the starboard bow, and before reaching it they could hear the fog horn.

On passing Cumbrae Light the master, who was on the bridge standing close to the telegraph on the starboard side, gave the order, or at any rate intended to give the order, for the vessel to be put upon a NNE course, and the order was repeated by the chief officer, who was standing near the helmsman on the port side of the bridge. Soon afterwards two green lights were observed ahead, upon which the helm of the *Rose* was starboarded for the purpose of passing them; and as soon as she had passed them, the original course was resumed. Presently a white light was seen on the port bow, and then a red light, which were taken to be the lights of a passing steamer; the vessel continued her course, passed the two lights, leaving them on the port hand; and shortly afterwards, and whilst still going at full speed, and making from 10 to 11 knots, she struck, and it was then found that the two lights were shore lights at Millport, and that instead of being, as they had supposed, well away to the west of the Great Cumbrae, they had run aground on the eastern shore of Millport Bay, not very far from Farland Point. There the vessel remained until yesterday, when we are told that she was got off, and was taken to Kames Bay, where she now is.

…No one contends for one moment that there has not been neglect somewhere in this case; the master says that it was the mate's fault, and the mate seemed to wish to attribute the casualty to an error in the compass. That there may have been more deviation in the compass than the master admits, is quite possible; but no conceivable amount of deviation could account for the casualty; for it is clear that the vessel could only have got where she did by steering an ENE course, instead of a NNE after passing the Cumbrae Light. That that course too was steered is clear also from the lights in Millport having been seen on the port side, as well as from the vessel's head having been, when she took the ground, ENE according to the master, and E by N according to the mate. And the only way in which it seems possible to account for an ENE course having been steered instead of a NNE course, is by supposing that the master may unconsciously, when he intended to give the order NNE have said ENE; and there is some ground for this supposition from what occurred when he was giving his evidence; he then told us that the order which he gave was ENE, but immediately corrected himself and said 'No, NNE.' …it appears to us that both the master and the mate are to blame for not having discovered the mistake before the vessel took the ground…

The **Holly** grounded going down the river from Sligo on the evening of 18 March 1884. The next tide fell off, but the succeeding one rose three feet higher than that on which the vessel left Sligo, and the wind drove her higher up on the shore. She finally refloated on 25 March and was able to proceed directly to Glasgow.

Three new ships were completed for the Laird company in 1884, all equipped with compound engines. The first to be launched was the small cargo steamer **Daisy** which was the last iron-hulled steamer ordered by the company. She was generally deployed on the Westport and Ballina west coast of Ireland roster. The next launch was that of **Thistle**, the first steel-hulled ship in the fleet and the first with electric lighting. She could accommodate 80 saloon passengers and 550 steerage passengers. The saloon was finely decorated in heavy drapes and dark wood panelling and there was a stained glass image of the company crest and a thistle on the dining room skylight. The dining room layout was of the traditional long table style. Her first arrival at Londonderry was on 11 July 1884, and was celebrated in the local newspapers with graphic descriptions of the 'fine new steamship **Thistle**'. The third launch was that of the **Elm** on 16 June 1884. She was destined for a nomadic existence, serving on whatever route she was needed, although she was often on the Morecambe to Londonderry service.

It was not long before the **Elm** was in trouble. On the morning of 11 September 1884 the **Elm**, on a voyage from Glasgow to Westport, went ashore on the east side of Rathlin Island at Castle Point. She had 50 passengers and some cargo on board; the passengers were taken ashore. After throwing 70 casks of oil overboard, the vessel backed off at high water in the afternoon. She then steamed to Ballycastle Bay, and afterwards to Church Bay, Rathlin, where she re-embarked her passengers. No one on board was injured and, as the **Elm** received no damage, she proceeded on her voyage to Westport.

The **Thistle** (1884), seen in her later life with white painted upperworks.

The passenger service to Portrush closed in favour of the new port at Coleraine in 1884. McNeill wrote in *Irish Passenger Steamship services*:

> In 1884 the deepening of the channel through the bar at the mouth of the Bann was completed and a new harbour was opened at Coleraine. In mid-August of that year the Laird service between Coleraine and Glasgow was revived and continued on a twice-weekly basis until the 1914-18 war. The steamers usually sailed from Glasgow on Monday and Thursday at 3 pm and returned from Coleraine, where the sailing times depended on the state of the tide, on Tuesday and Friday

afternoons. The fares were 10/- and 3/6d; passengers in a hurry had the option of joining or leaving the vessel at Greenock and making the journey between there and Glasgow by train at no additional charge.

The first steamer on the re-opened route was the *Fern*. She arrived at Coleraine in the early morning of Tuesday 19 August 1884 and disturbed late sleepers by sounding her 'brazen mouthed ordnance'. The next sailing was taken by the *Holly*, which left Glasgow on Thursday of the same week. She caused quite a stir when she arrived at Coleraine as at that time she was the largest steamer to have got up the Bann. In their first two months the vessels were advertised for cargo only but on the withdrawal of the Portrush-Glasgow steamers at the end of September the Coleraine-Glasgow ships began to carry passengers and continued to carry them until the service closed down about thirty years later.

The *Elm* (1884) became associated with the Morecambe services.

[Mike Walker collection]

Passengers for Portrush were sent ashore from the Coleraine boat by tender, and returned in the same way. Nevertheless the entry to the River Bann remained hazardous in northerly winds and the company inserted a clause in its Overall Notice to the effect that:

Cargo for Coleraine to be delivered at Portrush, in the event of the navigation of the River Bann being obstructed by weather or other cause, or of the vessel being unable from any cause to proceed to Coleraine; and the Company's undertaking for conveyance shall be deemed to be completed on Cargo being landed at Portrush, as if the cargo had been landed at Coleraine.

A similar clause dealt with the River Clyde and the termination of sailings from Ireland at Greenock rather than at Glasgow. Steamers could also miss the inward call at Greenock and go straight to Glasgow, presumably when lightly loaded or running well behind schedule. Any cargo for Greenock would be brought back later in the day by one of the outward-bound steamers. There is also a rather amusing indemnity for the carrier against the Act of God, the King's Enemies, Pirates, Restraints of Princes, Rulers, and People, or Legal Process…

A similar vessel to the *Elm*, which had been commissioned in 1884, was the *Gardenia* which arrived on duty in 1885. She was slightly smaller than the *Elm* and became closely associated with the Glasgow to Coleraine passenger and cargo service, calling off Portrush if required. Thus, in the course of three years, five new ships had been commissioned, one the *Daphne/Rose* having already been disposed of. This was a massive investment of well over £100,000 and reflected the buoyant trading conditions that the company then enjoyed. The export records for Londonderry show that in 1884 over 57,000 cattle, almost 15,000 sheep and more than 19,000 pigs were carried to ports in Britain by the Glasgow & Londonderry company and others.

Gardenia (1885) was designed for the Glasgow and Londonderry service.

[Mike Walker collection]

On 25 March 1885, the *Holly* was in contact with the *Clan MacDonald*, while on passage from Larne to Glasgow. Damage to both ships was slight. The *Fern* was stranded on 24 July 1885 on Lacken Brach, near Ballina. She was floated next morning at high water and continued her voyage. She did the same thing on Oyster Island near Sligo on 17 June 1886, when Captain Salter was able to get her afloat on the next high tide.

The success of the upgraded Glasgow to Dublin service was celebrated in 1885 by a new company name. On 1 October 1885 the rebranded Glasgow, Dublin & Londonderry Steam Packet Company Limited was incorporated under the management of Alex. A Laird & Company. Alexander Allan Laird was appointed manager of the new company on the 12 November 1885. Thereafter, most of the ships were registered as owned by that company although those involved with the Morecambe to Londonderry service generally remained under the ownership of Alex. A Laird & Company reflecting the close working relationship between Alex. A Laird & Company and the Midland Railway. Shortly before the incorporation the Steam Packet Company had applied the brand 'The Laird Line' to all its advertisements and literature so that in a short time the official title of the company was rarely alluded to. It had used the title The Laird's Line for some time in advertisements, which interestingly left out the old company title.

The *Azalea* suffered four separate minor groundings in the approaches to Sligo during July and August 1885. However, on the evening of 26 October 1887 the *Azalea* was proceeding down the Clyde at the start of another voyage to Sligo, when the steamer *Snipe*, destined for Port Glasgow with a cargo of iron ore, came into collision with her in Cartsdyke Bay at Greenock. The *Azalea* was caught on the port bow at the hawse-pipe, and had her plates ripped off for about 15 feet. One of the crew of the *Snipe* got on board the *Azalea*, which ran alongside Greenock Quay. The *Snipe*, owned by Alexander M Hay of Glasgow, made for Port Glasgow and the *Azalea* eventually sailed back up to Glasgow.

On 8 November 1886, the *Brier* spent just over 13 hours aground close to Blenwick Perch, inside Oyster Island, following her departure from Sligo. On 4 April 1887, the *Brier* was ashore at Rosses Bay, one mile from Londonderry, lying in a dangerous position. The 500 head of cattle that she had on board were taken off by tugs. She was refloated the following day and was able to reload her cattle and proceed to Morecambe undamaged.

A new passenger steamer, the *Ivy*, joined the fleet in 1888. She was the first steamer to be equipped with the relatively efficient triple expansion engine. She could maintain a sea speed of 13 knots. The livestock carried in the 'tween decks enjoyed for the first time fan assisted ventilation. This ship inaugurated the new Greenock to Portrush daylight service in 1890. The ship left Greenock at 11.30 am on Mondays and Fridays and returned from Portrush overnight.

At 7 pm on 19 November 1889, a collision occurred between the steamer *Morocco*, of Liverpool, and the *Holly*, in the River Liffey at Dublin. Damage was found to be slight and the *Holly* was able to resume her voyage to Glasgow next morning.

The little cargo steamer *Daisy* was sold in 1889.

Unloading the livestock, from *Passenger Ships of the Irish Sea* by Laurence Liddle

...the cattle trade, always substantial, rose to a peak in October and November...the loading and unloading of cattle was a noisy and often brutal business, calculated to turn the thoughts of anyone, with a vestige of feeling for animals, in the direction of vegetarianism. I did once hear a docker, assisting loading a consignment of lambs in Dublin, adjuring a colleague 'to go easy on the lambs', but usually the process was a mixture of shouts, heavy blows with stout sticks and prods with the metal tipped ends of the latter. Where two or three frightened beasts had, as not uncommonly happened, jammed themselves together in the narrow gangway they could be beaten across their faces to get them to back out.

CHAPTER 7

GROWING TRADE UP UNTIL 1904

Despite the self-satisfaction of rebranding and having arrived as part of the Irish Sea establishment, there were certain parties dependent on the Laird company who felt that they could do the whole thing better and at a cheaper price. Traders in Londonderry had long since complained that the Laird service to the Clyde was inadequate and expensive. In August 1888 they took matters into their own hands and launched a cheap, no frills service, with passage of a cow to Glasgow priced at 1/-, the same price as the steerage class fare. Shares were sold in a newly formed company, the Irish National Steamship Company, and two former railway paddle steamers were purchased and placed on the run. They bought the **Thomas Dugdale**, previously employed on the Fleetwood to Belfast route of the Lancashire & Yorkshire and London & North Western Railway Companies, and the **Stanley**, formerly on the Holyhead to Dublin crossing of the London & North Western Railway Company. The ships were registered under the ownership of the Irish National Steamship Company, Londonderry.

A hostile relationship developed at Londonderry between the new company and the Laird and the Burns 'establishment'. This all came to a head on Tuesday 6 September 1888, when there was a collision between the **Azalea** and the **Thomas Dugdale**, while racing up the Foyle to Londonderry. Passengers were alarmed but none was hurt in the incident. At the enquiry held a few days later in Londonderry it was reported that:

> From the evidence submitted on behalf of the Board of Trade, it appeared that the **Azalea** sailed from Glasgow, bound for Londonderry, on the evening of the 3rd of September 1888. Her crew consisted of twenty-four hands all told, and she was under the command of Mr John Hately, who holds a certificate of competency as master of a Home Trade passenger ship (No.102,038). Her draft of water was 10 ft. 1 in. forward, and 13 ft. 3 in. aft. She had on board a general cargo, and about fifty cabin and steerage passengers. The **Azalea** passed the Cloch Light at 9.55 p.m. on the 3rd of September, and arrived off Innishowen, at the entrance to Lough Foyle, about 6.20 on the morning of the 4th, proceeding up the Lough at full speed. The weather was fine and clear at the time, the wind being light from the westward. The ship reached Moville at 6.45, it being then broad daylight. There were on board some passengers for that place, and the steamer stopped to land them. She afterwards proceeded on her course towards Londonderry. In obedience to Bye-Law 44 of the Londonderry Port and Harbour Commissioners, she slowed her engines when passing a dredger that was in the Lough, and then proceeded on her course at full speed, putting on all pressure.

The **Thomas Dugdale** (1873) in her original London & North Western Railway livery before she was acquired by the Irish National Steamship Company.

[Mike Walker collection]

The Court was informed that between Redcastle and Whitecastle the paddle-steamer **Thomas Dugdale** was following the **Azalea**, pressing hard upon her, and gradually gaining ground; that the **Azalea** kept on her way on the port side of the channel, and that the two vessels were abreast of each other before she arrived at Ture Light; that the **Azalea** steamed close to the red perch on the south side of the channel; and that the **Thomas Dugdale** pressed the **Azalea** so closely on the starboard side, that her master, fearing a collision, stopped and reversed his engines to allow the **Thomas Dugdale** to pass. In less than half a minute from the time this was done the two ships collided. The **Thomas Dugdale**, however, continued at full speed, without stopping, until she reached Londonderry. The back-wash from the paddles of the **Thomas Dugdale** caused the **Azalea** to swing across the stream as soon as her engines were stopped; but when the road was clear she put her helm to starboard, turned ahead, and steamed into port.

On the evening of the 3rd of September 1888, the **Thomas Dugdale** left Glasgow for Londonderry with a general cargo, livestock, and fifty-nine passengers. She was commanded by Mr James Saultry, who holds a certificate of competency as master of a Home Trade passenger ship (No. 101,556), and had on board a crew of twenty-eight hands, all told. Her draft of water was 10 ft. 6 in. forward, and 11 ft. 7 in. aft. She passed the Cloch Light at 10.5 p.m. on the 3rd of September, and arrived off Innishowen Head at 6 a.m. on the following morning. It appears from the evidence given on the part of the owners and master of the **Thomas Dugdale** that she had some cargo and one passenger for Moville; but she did not stop to land them. The reason assigned for this was, that the passenger had no ticket or pass, and would not pay his fare.

Seeing the **Azalea** ahead, the master of the **Thomas Dugdale** proceeded onwards under a full head of steam, and at Redcastle he had arrived within 100 feet of her. At Drung Chapel he had got his bow abreast of the **Azalea's** quarter, and from this point every effort was made to get ahead of her. He did not even slow when he was passing the dredger, as required by Bye-Law 44 of the Londonderry Port and Harbour Commissioners. The race continued until the two ships arrived above Ture Light, when the **Azalea** sheered into the port sponson of the **Thomas Dugdale**, and after rebounding struck her a second blow abreast of the saloon bulwarks. The Azalea then stopped her engines and dropped astern, the **Thomas Dugdale** proceeding to Londonderry at full speed.

The two masters were 'severely reprimanded for 'rash navigation''.

In addition to sailing to Glasgow, the **Stanley** was also used on a service to Fleetwood. All went well until December 1888 when the steamer failed and was replaced, while repairs were put in hand, by the **St Magnus**, chartered from the North of Scotland and Orkney and Shetland Steam Navigation Company. A tense relationship continued between the Laird/Burns contingent and the Irish National Steamship Company throughout 1889, and much of 1890, when a steerage fare on all ships could be bought for just 1/-. Clearly this state of affairs could not continue with Laird and Burns losing money on both passengers and livestock. Discussions with the managing owner of the Irish National Steamship Company, Mr Michael M'Daid, led in September 1890 to Laird offering him £15,000 for the two old paddle steamers in return for the closure of the service. Shareholders in the Irish company were not best pleased to hear of the deal, claiming that they had not been consulted! However, both the Irish National Steamship Company and the Laird company would have shortly been facing insolvency if the deal had not been struck at that stage of this exceptionally prolonged rate war. The steerage fare, and that for cattle, increased fourfold overnight, back to the economically viable rate of 4/-.

The **Stanley** was resold for demolition, but the **Thomas Dugdale** was retained on the Londonderry cattle run and given the corporate name **Laurel**. She too was sold the following year, and was scrapped in 1892.

The final new major route was commenced in 1889 between Morecambe and Dublin, in conjunction with the Midland Railway, initially using the **Thistle**. Three sailings a week were given in either direction. This last piece of the network allowed a circular, four-ship, summer tour to be advertised: Morecambe to Dublin, Dublin to Glasgow, Glasgow to Londonderry and Londonderry back to Morecambe.

The *Glasgow Herald*, 15 September 1892 announced the end of an era:

> We regret to announce the death of Mr Alexander Allan Laird, senior partner of the well-known firm of Alexander A Laird & Co., shipowners, Glasgow, managers for the Glasgow, Dublin and Londonderry Steam Packet Company, which took place yesterday at his residence, Clyde View, Helensburgh. Deceased, who had reached the advanced age of 81, had been for a long time in failing health, and had been confined to bed for three months. Mr Laird was one of the oldest, if not the oldest, steamboat managers in Glasgow. He commenced his apprenticeship in the trade in 1824, under his father, who at that time had management of a line of steamers trading between the Clyde and Liverpool…

> After his father's death, Mr Laird became head of the firm of Alexander A Laird & Sons. Subsequently he was joined in partnership by Mr McConnell, and the firm became McConnell & Laird. On the death of Mr McConnell, 20 years ago, the firm was changed to its present form – Alexander A Laird & Co. Mr Laird's only child died a number of years

ago, but Mrs Laird survives her husband. Mr Laird has resided in Helensburgh for about 17 years, and was highly esteemed. He was for some time an Office Bearer in Park Free Church there, and was a liberal subscriber to the funds of the denomination.

In the summer of 1892 a new seasonal daylight service to Portrush was inaugurated by the **Azalea**, with departures from Glasgow via Gourock rather than the usual Broomielaw via Greenock. This was permissible because all cargo was handled at Glasgow, the Gourock stop was purely for passengers arriving and departing by train. Departures from Glasgow were scheduled for 9 am Tuesdays and Thursdays, 10.30 am on Saturdays, and returning from Portrush on Mondays, Wednesdays and Fridays at 9.30 am. The lay-over at Portrush allowed afternoon cruises to be carried out, including around Rathlin Island, to Red Bay and trips to Moville with a cruise along Lough Foyle. The service was operational only during the summer period.

In May 1893 the Londonderry to Fleetwood service was reinstated by the **Elm**, a service which for a while she shared with the **Ivy**. This route rapidly became a lucrative cattle run and was popular with saloon passengers travelling to England. It was operated in conjunction with the Lancashire & Yorkshire and the London & North Western railways.

Now that the devastating rate war on the Londonderry route had ended, funds were again built up ready to order a new passenger ship for the Dublin trade. In autumn 1892, Company Chairman Lewis McConnell, named after his father and one time partner of Alexander Laird, ordered a large steel-hulled passenger ship with coal-fired triple expansion engines from D & W Henderson & Company. The ship was launched in July 1893 and christened **Olive**. She was a fabulous ship capable of 15 knots, with accommodation for 100 saloon passengers and up to 1,000 deck passengers She started work later in the year on the Glasgow to Dublin service and was to be associated with this and the Dublin to Morecambe service for much of her early career.

Olive (1893) on trials in the Firth of Forth.

[Mike Walker collection]

The *Glasgow Herald* reported on 27 July 1893:

> Messrs David & William Henderson & Co. launched yesterday from their yard at Partick, a handsomely modelled steel screw steamer, which they have built to the order of the Glasgow, Dublin & Londonderry Steam Packet Company. The vessel has been built to Lloyd's highest class, but the scantlings are considerably in excess of their requirements. The regulations of the Board of Trade for a passenger certificate are also complied with. The vessel has been especially designed for the Irish Channel trade, and is fitted for the carriage of passengers, cattle and cargo. The principal dimensions are: length 260 feet, breadth 33 feet, depth 16 feet 6 inches.
>
> The accommodation for first class passengers, which is in the poop and deckhouse, consists of dining saloon and state-rooms, ladies' and gentlemen's cabins and smoke room. The dining saloon is a handsome apartment in polished birds' eye maple and oak and morocco leather. The state-rooms are large and well equipped. All the berths have wire mattresses, and the state-rooms communicate with the pantry by electric bells. Three of the state-rooms

are fitted in the deck house. They are of exceptional size, and are entered off a tiled passage. The companion and stairway are of polished teak, and are forward of these rooms in the same house. The smoking room is in a house at the fore end of the bridge deck. It is a large apartment fitted in walnut and upholstered in morocco. The floor is tiled and the place well lit and ventilated. The captain's room is in the same house, and is finished similarly to the smoking room. The officers' quarters are under the forecastle, where are also a room for the accommodation of the cattle dealers, and separate male and female steerages. The crew space is below this on the main deck.

The arrangements for the transport of cattle are most complete. All of the main deck that can be made use of without detriment to the passenger accommodation is fitted with cattle stalls, and the whole of the 'tween deck is similarly arranged. The ventilation of these spaces has been made a special feature. In addition to a thorough system of ordinary ventilation, a complete installation of artificial ventilation by means of fans on Messrs Lees, Anderson & Co's. principle has been introduced. By this arrangement the air in all the cattle spaces can be kept perfectly sweet. The vessel will have an installation of electric light by Messrs Wm Harie & Co., which will comprise the lighting of saloons, state-rooms, crew's and steerage quarters, engine room, decks and cargo hatches.

The arrangements for the rapid handling of cargo are necessarily very complete, and include three steam winches. The windlass is by Messrs Napier Bros., and the steam steering gear and winches are by Messrs Muir & Caldwell. Engines of high power are being fitted by the builders. The cylinders are 26 in., 42 in. and 63¼ in. diameter by 45 in. stroke. It is anticipated that the steamer will develop a high rate of speed. Throughout the construction of both hull and engines the work has been under the personal supervision of Mr Laing, the company's superintendent engineer. As the vessel left the ways she was named **Olive** by Miss McClennan, daughter of Lewis McClennan, esq, chairman of the company. The builders afterwards entertained the launching party at luncheon in the model room…

On Saturday evening, 19 May 1894, the **Fern** went aground inside the bar of the River Moy on her way to Ballina. She lay in a dangerous position with the sea washing over her and her rudder carried away. A tug and lighter were sent to assist but could not get alongside. The cargo was slowly discharged over the side at low water and carried across the strand by labourers. On 23 May, an attempt was made to move the ship under her own engines, then at low water the last of the cargo was discharged and ballast was taken on board to trim the ship to 8 feet fore and aft. The weather calmed and the ship was moved slightly, so that workmen could start to dig her out of the sand. On 30 May, the **Fern** was finally floated off using her anchors, which were laid out before her, and her own engines. She was anchored in the river and was found not to be taking in any significant amounts of water. The Clyde tug **Flying Huntress** was dispatched to tow the ship back to Glasgow for repairs.

Sail and steam at Coleraine: the Laird steamer **Lily** (1896) working cargo at the quay with a brigantine moored in the channel.

In 1895 and 1896 Blackwood & Gordon delivered two small passenger ships for the West of Ireland services. The *Daisy* was commissioned late in 1895 and the *Lily* followed at the end of 1896. Both ships were capable of 12 knots.

The *Glasgow Herald* reported on 20 September 1895:

> Messrs Blackwood & Gordon launched yesterday from their Castle Shipbuilding Yard, Port Glasgow, the ss *Daisy* to the order of the Glasgow, Dublin & Londonderry Steam Packet Company. The steamer is built of steel to Lloyd's highest class, and has dimensions: 190 ft. by 29 ft. by 15 ft. moulded to quarter deck, with a gross tonnage of about 560 tons. She is fitted for the carriage of passengers, cattle and cargo between this country and various Irish ports, and has an arrangement of water ballast that will bring her to a suitable trim under any conditions of loading. After the launch, which was successfully performed by Miss Turnbull, the steamer was brought into Messrs Blackwood & Gordon's wet dock to receive her engines: 15 in., 24 in. and 49 in. by 30 in stroke, and suitable boilers by the same builders. The steamer is fitted with all modern appliances, including a complete installation of electric light.

The *Glasgow Herald* again, in broadly similar vein, on 9 November 1896:

> Messrs Blackwood & Gordon launched on Saturday from their Castle Shipbuilding Yard at Port Glasgow, the ss *Lily*, to the order of the Glasgow, Dublin & Londonderry Steam Packet Company…She is fitted with passenger accommodation aft for 40 saloon passengers, also with steerage and deck accommodation, and is arranged for the carriage of cattle and cargo between this country and various Irish ports, and has water ballast fitted beneath the forward and aft holds sufficient to bring her to a suitable trim in any conditions of loading. After the launch, which was successfully performed by Miss Maclay, daughter of David T Maclay, secretary of the company…

It is notable that among those present at the launch of the *Lily* was Mr George Burns, then a director of the Glasgow, Dublin and Londonderry Steam Packet Company.

Lairdspool (1896) was commissioned as *Lily* for the service to Sligo and other west coast of Ireland ports.

[B & A Feilden]

Both the *Daisy* and *Lily* were associated with the various cattle runs and were used at Londonderry on the Liverpool service from time to time as well as on the Sligo, Ballina and Westport services. The other ship associated with the Liverpool service was the *Ivy*. However, the initial duties of the *Lily*, when she first appeared in November 1896 were on the Portrush daylight service from Greenock. An August 1896 newspaper advert for the Glasgow sailings listed the following departures from Greenock (generally two to three hours earlier from Glasgow):

> To Londonderry, every Monday, Tuesday, Thursday and Friday at 6.30 pm
> To Dublin, every Tuesday, Thursday and Saturday at 6.30 pm
> Daylight service to Portrush, every Tuesday and Thursday at 10 am
> To Coleraine, calling at or off Portrush, every Monday and Thursday at 6.20 pm
> To Sligo, every Wednesday at 4.40 pm and Saturday at 3 pm
> To Ballina, via Sligo every other Saturday at 3 pm
> To Westport, approximately alternate Tuesdays, pm

Fares:

	Cabin single/return	Steerage single/return
Londonderry	12/6d 20/-	4/-
Coleraine and Portrush	10/- 15/-	3/6d 6/-
Dublin	12/6d 20/-	6/- 8/-
Sligo and Westport	12/6d 20/-	6/- 8/-

In addition a special summertime overnight outwards and daytime return sailing was operated to Portrush on Fridays at 10.10 pm returning Monday morning at 10 am. For this a weekend trip to the Giant's Causeway was offered '1st class 42/-, including, Rail, Steamer, Tram to Causeway, Board Lodging in the Causeway Hotel, Saturday till Monday.'

In addition there was a regular service between Fleetwood and Belfast, in collaboration with the London & North Western Railway, and between Dublin and Morecambe and Londonderry and Morecambe, calling at Larne, as well as from Liverpool to Londonderry and Fleetwood to Londonderry. This complex network required careful rostering of the fleet and scheduled cover for overhaul periods. The ships were broadly of two types, the larger passenger, livestock and cargo carriers on the Dublin and Morecambe routes: **Azalea**, **Cedar**, **Shamrock** and **Olive**; and the smaller ships designed for the north and west Irish trades: **Fern**, the oldest ship in the fleet dating from 1871, **Brier**, **Thistle**, **Elm**, **Ivy**, **Daisy** and **Lily**. The ships were nevertheless interchangeable and it was not uncommon to see the **Azalea** and **Cedar** at Londonderry and even, on occasion, the **Shamrock** would move off her Dublin to Glasgow roster to take the Londonderry service from Glasgow.

Not surprisingly ships were occasionally chartered in to cover for refits and accidents. Minor strandings and collisions still happened, an inevitability given the changeable weather conditions and confined waterways at the entrance to many of the ports served by the company. Ships were also chartered out, the **Elm** for a lengthy period to G & J Burns, complete with that company's black funnel.

Despite the small recession of the late 1890s, the company continued to prosper and to offer an excellent service to shippers and passengers at acceptable rates and fares on both sides of the Irish Sea. The Managing Director, William MacConnell (junior) must have been well pleased with the performance of his company. In addition, its close liaison with both the Midland Railway and the London & North Western Railway continued to bear fruit, the former conducted through Alexander A Laird & Company, the latter directly with the Glasgow, Dublin & Londonderry company. Services essentially remained unchanged, although passenger carrying to Coleraine was discontinued in 1900.

The **Fern** grounded at the entrance to the Bann River on 20 January 1897, and was beached in the river with the fore hold full of water. Two days later she was refloated and taken to Coleraine for repairs. She arrived at Port Glasgow to be slipped on 25 January. On 2 May 1897, the **Fern** went ashore on the Channel side of Holy Island, near Lamlash, on arriving in the Firth of Clyde from Coleraine. The steamer **Shamrock**, coming in from Londonderry, took the passengers off and conveyed them to Glasgow. Next day it was reported that the forehold was full of water. There was a hole on the port side some 16 feet by 3 feet and another on the starboard side 5 feet by 3 feet. The tug **Flying Wizard** took off the cattle and discharged them at Glasgow. The remaining cargo was discharged into the puffers **Polarlight** and **Sealight** by 4 May. The **Fern** was finally refloated on 15 May and taken to Lamlash Bay where she was beached. Once patched up, she was refloated on 17 May and sailed under her own steam to Port Glasgow for permanent repairs. The **Fern** was sold shortly afterwards.

A new steamer for the west of Ireland service was commissioned in summer 1900. This was another **Fern**, a modest steamer designed for the Sligo trade with calls at Westport and Ballina. Built at Troon by the Ailsa Shipbuilding Company, she was equipped with triple expansion machinery built by Dunsmuir & Jackson at Govan with cylinders of 16 inch, 26½ inch and 43 inch diameter with a stroke of 30 inches.

The popularity of the Glasgow and Gourock to Portrush day sailings during the summer season was such that, in the summer of 1900, a two-ship daily service was inaugurated and a one ship service maintained in the off season.

Perhaps the most celebrated of all the Laird ships was the **Rose**, which was built by A & J Inglis at Pointhouse and completed in 1902. So appropriately designed was she that she satisfied the needs of the Irish trades for the next 47 years. She was a big ship of 1,151 tons gross equipped with triple expansion engines which gave her a sea speed of 15 knots. She had a deadweight capacity of 650 tons including 118 tons coal bunkers. Her length to breadth ratio was 6.9, being 3 feet broader than that of the **Olive** which had a length to breadth ratio of 7.9. This width, coupled with the fitting of bilge keels, made the new ship particularly stable, a feature that endeared the **Rose** to her regular passengers. She had a long bridge deck, beneath which was accommodation for 100 first class passengers, while an additional 40 passengers could be berthed on sofa beds. Up to 700 steerage passengers could be accommodated in the poop, where there were numerous sparred seats available for them. Steerage class toilets were also fitted in the forward 'tween deck so that, in season, people could move into the cattle spaces which, hopefully, had been cleaned down and appropriately furnished.

Fern (1899) was deployed initially on the Clyde service to Sligo.

[Mike Walker collection]

The ***Rose*** (1902) in Burns & Laird Lines Limited colours heading down the Clyde with a full load of passengers on board.

Charles Waine wrote:

> She was typical of the Glasgow built and owned vessels of the period serving Ireland. She had two decks and a third weather deck, which had deep wells, so though she appeared flush decked she was described in Lloyd's Register as having poop, bridge and forecastle, although with the portable decking covering the forward wells when at sea, the deck became continuous for practical purposes. Side doors allowed cattle to walk aboard and down ramps in the hatchways to the lower deck while horses were accommodated on the upper deck.

She started work in late summer on the Dublin and Glasgow route, displacing the **Azalea**, which took up service to Londonderry. She quickly became popular with her first class clientele, many of whom waited until it was her turn to sail before travelling. The **Gardenia** was sold in 1904, having been associated mainly with the Portrush and Coleraine routes to Glasgow, but latterly employed at Londonderry.

The **Fern**, under the command of Captain Campbell, grounded half mile east of Banmouth Coastguard Station in the evening of 7 October 1903. She had just left Coleraine for Glasgow. The **Fern** lay on the north side of the river about a mile and a half inside the bar, and was out of the channel; she had grounded one hour after high water on a very high tide and on 9 October started to land all the livestock and lighten ship. She was still on the ground on 21 October after several attempts had failed to get her off, but then at 7 pm, on a spring tide, she was finally released, apparently little damaged.

November 1903 sailing schedule for the Morecambe to Dublin service.

On 13 August 1904 the Midland Railway opened its new dock at Heysham with a departure of its new steamer **Londonderry** to Douglas. The Belfast service started on 1 September, and the Laird Line schedules switched from Morecambe to Heysham that same week. The Midland Railway introduced four new steamers for Heysham, three for service to Belfast and one, the triple screw turbine steamer **Manxman**, to serve the Isle of Man. The Belfast steamers, **Antrim**, **Donegal** and **Londonderry**, the latter with direct turbine machinery and triple screws, were scheduled to depart at 10.00 pm, after the arrival of the London train. The Laird berth was on the landward side of the Railway steamers' berth, the latter being adjacent to the newly constructed railway station. Although the approach channel to the harbour entrance was not an easy one to navigate, it was far superior to the approach to Morecambe with its ever-shifting sandbanks.

Dick Clague in his history of Heysham Port offers:

> Laird Line switched their Morecambe services to Heysham immediately the new port opened – daily to Dublin and twice weekly to Londonderry. Unlike the Midland Railway, Laird Line used many different ships on what were

Olive (1893) alongside at Heysham with one of the Midland Railway 'Dukes' lying ahead of her.

[Mike Walker collection]

A remarkable photograph of the *Olive* (1893) at the Laird berth at Heysham Harbour dated 21 September 1904, just three weeks after the Harbour had been opened. One of the Midland Railway ships lies beyond her at the railway berth and Alexandra Towing Company's tug *Langton* is in the foreground.

[Mike Walker collection]

referred to as their 'outport services' from Heysham. In 1905 the *Olive* and *Thistle* were usually on the Dublin run (which the *Shamrock* had opened the previous year) and the *Brier* and *Azalea* served Londonderry with a call en route at Portrush. By 1909 the *Hazel* (later sold to the Steam Packet as the *Mona*) appeared on their rosters.

The cross-channel trade from Londonderry reached its greatest activity about 1904, while the Lancashire and Yorkshire and London and North-Western Railways' service to Fleetwood was maintained. Their railway steamers left Londonderry on Tuesday and Friday at 4 pm and caught the 6 am boat-express due in London at 11 am the next day. The return service left London at 5.30 p.m. on Wednesday and Saturday, and the steamer reached Derry at 11 o'clock next morning. The saloon and third class return fare was 45/-. At this period there were two, if not three, sailings weekly to Liverpool, two to Heysham, and six to Glasgow, all carrying passengers.

John Kennedy in *The History of Steam Navigation*, summarises the former Morecambe, Fleetwood and Liverpool trades in 1903:

In connection with the Midland Railway Company of England a service of powerful steamers is maintained between Morecambe and Dublin, the steamers sailing from the respective ports on alternate days, and making the passage in about 10 hours. Early next year, it is proposed to transfer the service from Morecambe to Heysham, and to maintain daily sailings to and from the latter port and Dublin. The Laird steamers also sail in connection with the same railway company from Morecambe to Londonderry every Tuesday and Saturday, returning from Londonderry every Monday and Thursday.

From Fleetwood, in connection with the Lancashire and Yorkshire, and London and North Western Railway Companies, a weekly service had been maintained for many years by the company's steamers between Fleetwood and Londonderry, but in September, 1903, Messrs. Laird & Co. retired from this service.

From Liverpool also, steam communication is maintained with Larne, Coleraine and Westport. The fixed sailings are once a week from each port, but extra steamers are dispatched according to the requirements of the trade.

The company's fleet at the present date (1903) consists of 12 first-class powerful steamships, having an aggregate gross tonnage of 9,164 ton, and named as follows [gross tonnage in brackets]:

Azalea (748)	*Elm* (521)	*Olive* (1141)
Brier (728)	*Fern* (503)	*Rose* (1363)
Cedar (750)	*Gardenia* (491)	*Shamrock* (804)
Daisy (565)	*Lily* (668)	*Thistle* (822)

The *Brier* (1882) coming through the Heads at Heysham in May 1909 inbound from Londonderry.

[Mike Walker collection]

The *Azalea* (1878) alongside at Heysham.

[Mike Walker collection]

CHAPTER 8

LAIRD LINE LIMITED

The cumbersome title Glasgow, Dublin & Londonderry Steam Packet Company finally gave way to the more popular name Laird Line Limited; the company name was changed by special resolution on 15 January 1907.

The next new ship to join the company, the fast passenger and cargo ship **Hazel**, was a rather special one, as described in the *Glasgow Herald*, Monday 15 April 1907:

> Ardrossan to Portrush Laird Line Daylight Service: A new twin screw steamer for the Laird Line (Limited), Glasgow, was launched on Saturday by the Fairfield Shipbuilding & Engineering Company (Limited), Govan. This vessel, which was named **Hazel** by Mrs Lewis MacLellan, wife of one of the Directors of the Laird Line, is intended to inaugurate a new 'daylight' service between Ardrossan and Portrush, going and returning the same day, and doing the single run in about four hours. She is expected to be on the station on the 15th of June. The new service will make it possible for passengers to go to Portrush and come back within ten hours, and it will also open direct connection between Scotland and such important towns as Ballycastle, Coleraine, Ballymena, Ballymoney and Cookston. When the **Hazel** begins to run travellers will be able to leave the east coast of Scotland in the morning and reach places in the west coast of Ireland the same afternoon.
>
> The general dimensions of the new steamer are: length 258 feet, breadth 36 feet, depth 23 feet 9 inches. There are four decks – the lower, the main, the bridge and the boat deck. The first class accommodation is situated forward from amidships, the second class aft. The bridge deck amidships is reserved for the exclusive use of first class passengers. For these there has been provided a dining saloon seating eighty persons, a ladies' cabin, ladies' and gentlemen's toilet rooms, tea room with fruit and confectionary room. The dining saloon is neatly finished in mahogany with satinwood inlay panels; and the tables are arranged for small parties in the manner adopted by modern hotels. A special feature of the saloon is the two after bays which are arranged so that they can be reserved as 'cosy corners'. The tea room is neatly finished in light oak, with tulip wood inlay panels. It will be fitted in a thoroughly up-to-date manner, and will be run on city lines. On the boat deck there is a large smoke room finished in oak with red marble tables, and there is also an upholstered deck lounge, as well as private state rooms for first class passengers. For second class passengers there is a large dining saloon on the main deck. The whole of the midship part of the main deck can be used as a promenade for second class passengers or as a shelter in wet weather.

Hazel (1907) embarking passengers at Portrush.

[DP World]

The Laird Line brochure for 1908 advertising summer tours in Ireland based on the regular Glasgow steamer services.

The propelling machinery consists of two sets of triple expansion engines, each set having four cylinders and four cranks and being balanced on the Yarrow-Schlick-Tweedy system so as to reduce to a minimum the vibration usually caused by reciprocating engines. Steam is supplied by four single-ended boilers of the cylindrical return-tube type, arranged in two boiler rooms. The vessel has been constructed to Board of Trade requirements and under Lloyd's Special Survey, and also under the supervision of Mr James Maxton, consulting engineer, Belfast, along with the owner's superintending engineer, Mr W Robertson. The launching weight of the vessel was 900 tons, and the time taken in going down the ways 45 seconds.

Direct acting turbines had been considered for the new ship but were rejected because of her light draft and short length which were needed to get safely into Portrush. Indeed her length to breadth ratio was just 7.2. Nevertheless, her two triple expansion steam engines driving twin shafts gave her an impressive sea speed of 19 knots. The **Hazel** is also notable in that she had almost no cabin accommodation, being designed principally as a day boat. She did start on the Ardrossan service on 15 June, having given a cruise around Arran on 11 June for 250 invited guests and the press. She generally left Ardrossan at 9.45 am and Portrush at 3.30 pm. The inauguration of the new daylight service allowed the Gourock service to terminate, releasing the two ships that had been needed to carry this out, each ship having taken two days for the return trip. The new service from Ardrossan quickly became very popular, with connecting trains allowing easy access to and from Glasgow.

The Ayr Steamship Company ran a daily steamer to Belfast leaving Ayr at 10 am with a connecting train from Glasgow leaving at 7 am. The company also ran to Larne on Mondays, Wednesdays and Fridays with a departure time from Ayr of 7.15 pm.

Charles Waine wrote:

> A small but busy company formed in 1876 to run steamer services between Ayr and Larne, Belfast and Campbeltown, with sailings also between Glasgow and Barrow-in-Furness. Weekly service to Dublin was added in 1900. Iron from Ayrshire furnaces was the principal cargo, hence the service to Barrow. The company was originally managed by P Barr & Company and from 1883 by D Rowan (later Rowan & Bain). New and second-hand steamers were acquired, the latter from the fleet of the Clyde Shipping Company, well suited to the passenger, livestock and general cargo trades. Operations were not confined to the regular sailings and the **Mona** was lost on a voyage from Mull to Stranraer with livestock in March 1908.

Turnberry (1889) was one of a number of ships acquired in the takeover of the Ayr Steam Shipping Company in 1908 seen here as Burns & Laird Line's **Lairdsheather**.

[B & A Feilden]

Not surprising then that this company caught the eye of Lewis McConnell and his directors as a potential acquisition of the Laird Line. Following negotiations between the two companies, the majority ownership of the Ayr Steam Shipping Company was purchased by the Laird Line with effect from May 1908. The company was allowed to retain its own identity and manage its own affairs, although from then on its business was carefully scrutinised by the new parent company. Its ships, however, had to adopt the Laird Line funnel colours. At the time the Ayr company was taken over by the Laird Line, it had just lost the **Mona**, but had the steamers **Turnberry**, **Merrick** and **Dunure** in service. The **Merrick** was wrecked on 29 December 1908, near Lendalfoot south west of Girvan on a voyage from Ayr to Girvan with coal. She was a cargo steamer with a compound engine, built as the **Amsterdam** in 1878 for James Rankine, of Glasgow. She had been bought by the Ayr company in 1903. The other two ships had passenger accommodation. The **Dunure** was none other than Laird's former steamer **Cedar**, sister of **Azalea**, dating from 1878, and which had been sold to the Ayr company in 1906. The **Turnberry** had been acquired in 1899 from C F Leach, of London, for whom she had been built in 1889 as **Spindrift** and equipped with triple expansion steam machinery.

In May 1909 a new steamer with the name **Rowan** joined the Laird Line's fleet. She was the biggest ship in the fleet and measured 1,493 tons gross. Unlike the smaller but faster **Hazel** on the Portrush day service, the **Rowan** had a single screw and a conventional triple expansion engine, capable of giving the ship a sea speed of 16 knots. She was distinctive as she carried a large funnel cowl that was intended to keep coal smuts off the deck. The **Rowan** was designed for the Glasgow to Dublin service, for which she had a large deadweight and extensive passenger accommodation. Her arrival into service allowed the **Elm** to be sold early in 1910. Brian Patton wrote:

> Accommodation for passengers travelling cabin comprised a small lounge on the upper deck, a dining saloon forward on the promenade deck, a smoke room aft on the same deck and cabins on that and the main deck. For those in steerage, there was a ladies' lounge in the poop deckhouse and a general room below.

On 25 November 1909 the **Hazel** was in collision with the **Minard Castle** near Renfrew on the Clyde. The **Hazel** was inbound from Londonderry and the **Minard Castle** was setting off for Loch Fyne ports. The *Glasgow Herald*, 26 November 1909 reported:

> A collision took place early yesterday morning on the River Clyde near Renfrew. The Lochfyne & Glasgow Steam Packet Company's **Minard Castle** was going down channel on one of her regular runs to Lochfyne ports, and the Laird Line steamer **Hazel** was coming up on her inward voyage from her usual Londonderry trip. The two vessels met at Renfrew, and owing to a haze which prevailed just before daybreak they did not see each other in time to prevent their bows striking. They struck with so much force that their bows were very badly damaged. The damage, however, was in each case all above watermark, and although there was at first some alarm on board it was soon seen that both vessels remained absolutely seaworthy and were not in any further danger. The damage to the **Minard Castle** was, however, so serious that she could not proceed on her voyage, and she was therefore put back and berthed in Kingston Dock to discharge her cargo preparatory to being repaired. The **Hazel** proceeded to the Laird Line berths. She will also require repairs before she is again fit for use.

During the summer of 1912 the Ayr company steamer **Dunure** was used to provide a number of Clyde cruises with departures from Ayr and Troon. These were repeated in 1913 and 1914 during the Glasgow Fair Week and on selected weekends using the steamer **Lily**.

The last new ship to be ordered by the Laird Line was the **Maple**. She was considerably smaller than the **Rowan**, but she was nevertheless a useful ship for the Londonderry service on which she replaced the **Azalea**, in summer 1914, allowing the latter to be sold. The **Maple** was equipped with the now standard triple expansion steam engine with cylinders of 28 inches, 45 inches and 72 inches with a stroke of 42 inches and was fit for 15½ knots. She provided 112 comfortable first class berths amidships with second class accommodation plus some deck class space under the poop (the terms saloon and steerage having been superseded). Both **Rowan** and **Maple** had the usual arrangement for loading and unloading livestock onto the lower and main decks.

Britain declared war on Germany shortly after the **Maple** had been delivered. Many thought the war would be over by Christmas and business tended to carry on as normal. However, the **Hazel** was withdrawn from her daily service to and from Portrush in August 1914, but other services were maintained broadly to schedule. On 14 November the **Hazel** and the **Rowan** were requisitioned as armed boarding vessels and stores ships to become HMS **Hazel** and HMS **Rowan** respectively.

Commander Charles P Wilson RNR was appointed to take command of HMS **Hazel** and engineer William Oswald RNR appointed as chief engineer. She otherwise retained much of her civilian crew. Throughout 1915 she was on duty as an armed boarding vessel on the approaches to the Irish Sea, a tedious job with much waiting about in rough seas, interspersed by short bursts of excitement. She was dry-docked in May 1916, preparatory to going to the Mediterranean where she served at Mudros, in the North Aegean, Suda Bay, in Crete, and Port Laki. She was used for landing troops on various beaches and came under air attack during these duties. In May and June 1918 she was

Maple (1914) was built for the Londonderry service as a replacement for *Azalea*.

[Mike Walker collection]

The final profile of *Maple* (1914) post 1936, complete with forward observation lounge and with the new name *Lairdsglen*.

[Mike Walker collection]

Azalea (1878) with a full complement of deck passengers.

[Mike Walker collection]

on duty sailing between Milos and Mudros, following a visit to Port Said with HMS *Colne* as escort. HMS *Hazel* left Mudros for home, arriving at Malta just after Christmas and eventually reaching the Clyde in January1919. She was decommissioned on 18 February 1919 and returned to her owners in May 1919 after the removal of her two 76/40 guns and deck platforms, and partial refurbishment.

HMS *Rowan* was commissioned as an armed boarding vessel on 30 November 1914 under Lieutenant Commander Walter Wilkes RNR with John McIsaac RNR appointed as Chief Engineer Officer. After duty in home waters, she went to the Mediterranean, calling at Naples for stores in November 1915 and on to Salonika. As with HMS *Hazel*, HMS *Rowan* spent much of her time serving at Mudros, and later also Salamis, where she tended to berth alongside the battleship HMS *Exmouth* for stores. HMS *Rowan* was not decommissioned until 10 April 1920 and was returned to her owners on 15 June 1920. She was off commission between 7 July 1917 and 11 July 1918 when she was used as a transport in home waters.

HMS *Rowan* at Malta in April 1919 after assisting at Gallipoli.

As the war progressed, the Laird Line came under increasing pressure, while restrictions on sailings and slow turn-rounds at docks played havoc with any attempt at regular departure schedules. For a while, the **Maple** left her Londonderry roster for the Heysham to Dublin route in the winter of 1915-16, and several of the ships were to be seen on otherwise unfamiliar services.

Rose and **Maple** were allocated to trooping duties in 1917 and 1918. They worked alongside each other at one stage on the Taranto to Itea ferry from Italy to Greece; other duties included occasional trips to Alexandria, Piraeus and elsewhere in the eastern Mediterranean. The **Maple** also served as a transport nearer to home earlier on in the war. Many of the older ships, including the stalwart **Olive**, remained in commercial service throughout the war.

Two ships were lost, the *Daisy* while attempting to enter the Bann on the approach to Coleraine and the **Fern** to enemy action. Whilst attempting to cross the bar to enter the River Bann, a heavy sea lifted the stern of the *Daisy* forcing her bow to strike down on the bar, and she ended up alongside the west pier with her bow facing towards the sea. The passengers and crew landed on the west pier using the ship's boats. *Lloyd's List* 24 February 1915, reported:

Portrush 23 February, 2.10 pm. Laird steamer *Daisy*, of Glasgow, from Glasgow to Coleraine, with passengers and general cargo, grounded in River Bann, Bannmouth at noon today; in dangerous position.

Portrush 23 February, 4.18 pm. Steamer *Daisy* ashore on west pier, Bannmouth. Crew and passengers landed at Castlerock, rudder damaged. Making quantity of water; position dangerous.

The following day the ship was described as lying in a critical position with her bows touching the west pier and her aft hold flooded. However, no salvage attempt could be made until the salvage company arrived on 27 February. Neither salvage vessel nor pumps were available and by 1 March the *Daisy* had sunk in the river:

Glasgow 1 March, 3.09 pm. Glasgow Salvage Association have following telegram sent by their surveyor from Castlerock: Steamer *Daisy*. Weather stormy; vessel now listing badly. Sea rolling completely over her fore and aft. Some cargo and deck fittings washed ashore. Have advised receiver for which. Salvage steamer not arrived.

Glasgow 3 March. Steamer *Daisy*. Salvage steamer arrived, operations commenced, weather improved, swell on bar; afraid little progress today.

Glasgow 5 March. Following telegram received from surveyor re steamer *Daisy*. Regret break in weather; heavy swell on bar also five knot current in river running out preventing all attempts to salvage ship or cargo. I recommend that operations be suspended until April or May, salvors same opinion. Surveyor has been telegraphed act on his recommendation sending back salvage expedition and stopping all expenses. Glasgow Salvage Association.

The contract to dismantle the wreck of the *Daisy* was awarded to Robert Lemon, of Peel, Isle of Man, who arrived that month with his salvage vessel **Diver Lass**. He was still working on the wreck in April 1916 when the **Corsewall**, of Glasgow, also stranded on the west pier. During an attempt to get a line on board, Mr. Lemon's working boat was destroyed and he was thrown into the sea, but survived by climbing onto the rigging of the wreck on which he was working.

The second wartime casualty was the **Fern**, whilst on a routine voyage from Dublin to Heysham. On 22 April 1918, the **Fern**, just two hours out from Dublin, was struck by a torpedo fired by submarine U104; the **Fern** sank within five minutes. She had 22 crew and 15 passengers on board; only 23 of the ship's complement survived. The **Fern** still lies at a depth of 79 metres at the south-eastern end of the Lambay Deep, which is one of the deepest parts of the Irish Sea, reaching depths of more than 130 metres.

The previous month the **Lily** had been in collision with the Gem Line steamer **Diamond**, dating from 1896. **Lily** was on a routine voyage from Londonderry to Glasgow. On 13 March 1918 at 12.40 pm the **Lily** struck the **Diamond**, loaded with steel plates, causing her to sink in just a few minutes. The **Diamond** had been on passage from Cardiff to Londonderry and the collision occurred off Rathlin Island, one mile north west of Altacarry Lighthouse. The master and three others were rescued, but nine of the crew were lost. The **Lily** had her bows damaged and the forepeak filled with water to a height of 7 feet. The next compartment also leaked, but the pumps kept the water under control.

The **Maple** suffered a fire in her fore hold while on passage from Glasgow to Dublin on 3 March 1915. She was able to reach Dublin safely, getting the fire under control, but with considerable damage caused in the hold. The 'tween deck beams and deck plating at the after end of the hold were buckled across the whole width of the ship and the spar ceiling and bulkhead sheathing were burnt away. The **Maple** was sent to Troon for repairs. The whole after end of the 'tween decks had to be lifted for the work to take place, and a side shell plate had to be cut away to provide access for the beam repairs.

One new ship was acquired in August 1915 and given the name **Broom**. She had been built in 1904 as the **James Crombie** for the Aberdeen, Leith & Moray Firth Steam Shipping Company Limited. That company had been acquired by M Langlands & Sons in 1914. However, the **James Crombie** became surplus to requirements; being of limited deadweight, she could not easily be integrated with the normal Langlands' services. The Laird Line, however, was short of tonnage, with ships either requisitioned or away on trooping duties, and viewed the Langlands' ship as a useful stopgap to see the war through. Stopgap or not, she remained in service until 1949, latterly under the management of Coast Lines Limited.

The **Shamrock** went aground on 1 June 1917 in dense fog almost at low tide four miles south of Skipness on the Mull of Kintyre. She was on passage from Belfast to Ayr. She lay almost parallel with the shore, with her stern high, bow down, and at mid-tide the main and fore holds were flooded. At high tide the water was almost flush with the deck. Some 75 cattle were landed safely, a further 22 were drowned. The **Shamrock** was eventually refloated four weeks later and was taken to Ardrossan Harbour where she was beached for inspection. A further four weeks lapsed before she was able to get into dry dock for repairs, mainly to bottom damage.

Following the Armistice on 11 November 1918, the merchant service strove to get back to business as usual. However, it was a slow start and commerce and trade were at a low ebb. This was especially so in the Irish trades with civil disquiet in Ireland discouraging commerce and certainly closing the door on any likely tourist trade. So, when the **Hazel** was released by the Admiralty on 11 May 1919, her owners had already made the decision that her pre-war express daily return to Portrush was no longer viable. They were able to arrange the sale of the vessel to the Isle of Man Steam Packet Company to take place just ten days after her return – on 21 May. As the **Mona**, she provided a useful back-up and winter unit for that company until sold for demolition in 1938.

The Isle of Man Steam Packet Company's **Mona** (1907), formerly the **Hazel** built for the Ardrossan to Portrush day service and photographed in the Mersey by Basil Feilden.

[Mike Walker collection]

Coast Lines Limited was formed in 1919 by the amalgamation of three major coasting interests: F H Powell & Company, John Bacon Limited and Samuel Hough Limited. Alfred Read, Chairman of newly created Coast Lines Limited, and a part of the expansive Royal Mail Group, had access to Royal Mail money. Coast Lines' ownership was spread across the Royal Mail Group, between Royal Mail Steam Packet Company, Elder Dempster, Union-Castle Line, H & W Nelson, Lamport & Holt and D MacIver. The cash available to Read was to enable expansion of his network, partly as a feeder service for Royal Mail Group's deep sea ships. The Laird Line Limited's sphere of operations in the Irish Sea, north and west coast of Ireland were entirely complementary to those which Alfred Read already owned, so an approach was made directly to Chairman William MacLellan (grandson of Lewis MacLellan, and a founder of what was later to become the Laird Line). A deal was struck and a price of £933,000 was agreed between the two men for the goodwill, assets and staff of the Laird Line, with a promise that it could keep its identity for the immediate future. The effective date was 7 October 1919. The price was low for what Coast Lines was getting, but bearing in mind the poor trading conditions in the Irish trades at that time and the equally poor prospects for the foreseeable future, it was probably a fair one. Outwardly nothing changed and the company continued as before, save for a Coast Lines placement into the company board room.

Being now part of a larger network of ships and services, the **Maple** was deployed on the Fishguard to Cork Service of the City of Cork Steam Packet Company from December 1920 to May 1921. She was replaced on that service on 31 May by the Belfast Steamship Company's **Classic**.

For the next couple of years the Laird Line was left pretty much to its own devices. In 1921, the Ayr Steam Shipping Company ceased to trade and its ships, **Turnberry** and **Dunure**, and other assets were absorbed into Laird Line Limited. In 1921 Coast Lines transferred two elderly and obsolescent passenger units from its subsidiary Belfast Steam Ship Company Limited, which it had also acquired in 1919, to Laird Line Limited. These were the **Comic**, dating from 1896, and the **Logic**, completed in 1897, each with a book valuation of £15,000. They were given Ayr Steam Shipping Company style names, respectively **Cairnsmore** and **Culzean**, and placed on the Ayr to Belfast and Ayr to Larne services of that former company.

The year 1921, however, is better remembered for the '**Rowan** Tragedy' that occurred on 9 October 1921. The Glasgow Herald 10 October 1921, reported:

> By a double collision in the Irish Channel early yesterday morning during fog a number of lives, estimated at 21, were lost. The Glasgow steamer **Rowan** on a voyage to Dublin, was struck first by the American steamer **West Camack**, and then by the Glasgow steamer **Clan Malcolm**. She was seriously damaged by the first blow, but would, it is thought, have continued afloat until all those aboard had been saved. Ten minutes later, however, she was struck right amidships and practically cut in two, sinking within three minutes. Of those on board 77 were rescued, but two died afterwards. The exact number of lives lost has not yet been ascertained but there are unaccounted for 13 members of the crew including the captain, and also probably eight passengers.

> The **Rowan** left Glasgow on Saturday evening for Dublin on her usual run with passengers and cargo. She cleared from Greenock a little after seven o'clock, having been delayed for about an hour waiting for the arrival by train from Glasgow of the members of the Southern Syncopated Orchestra, who were completing an engagement at the Lyric theatre, Glasgow, and could not join the boat before she left her berth. When five hours out from Greenock, the **Rowan** ran into banks of floating fog off Corsewall Point. While she was passing through the fog she was speeding at 10 to 12 knots, somewhat lower than her usual service speed, when she was struck on the stern by the American steamer **West Camack**, inward bound from the United States with general cargo. This collision took place at 12.10 am – did not cause very serious damage, and Captain Burns of the **Rowan**, at once gave orders for all the necessary precautions to be taken. Lifebelts were being served out and arrangements were being made for the lowering away of

Culzean (1898) was transferred from the Belfast Steamship Company Limited as a replacement for the Ayr services in 1921.

the boats. Everything was proceeding in good order and there was every likelihood that all on board would be saved before the vessel was in danger of going down. Suddenly, however (ten minutes after the first collision), the Glasgow steamer **Clan Malcolm**, outward bound for Natal, Delagoa Bay and Mauritius, appeared out of the fog and crashed into the **Rowan** almost at right angles, a little forward of amidships on the starboard side. The **Rowan** was practically cut in two, and sank within less than three minutes, giving those still below little chance of being saved. Almost immediately after the vessel sank the fog lifted.

Rescue ships at the scene included the destroyer HMS **Wrestler** and the Burns steamer **Woodcock**. At the subsequent Inquiry Captain Donald Burns was posthumously chastised for not having reduced his speed even more than he did and for not sounding his whistle as he proceeded in the fog. Once struck by the **West Camack** he was then also reprimanded for not ringing the ship's bell to warn of a stationary vessel in fog.

The incident was reported globally and attracted great interest in America because of the black American jazz orchestra that was on board, the drummer of which, Pete Robinson, was lost. The *New York Times* 10 October 1921, included the statement:

The hero of the tragedy was Egbert E Thompson, leader of the orchestra, who served in France in the war with the 'Buffaloes' infantry. He was carried down by the ship but struggled to the surface and swam to a life raft, onto which he dragged men, women and children from the water…

After the surviving men had returned to London to purchase new instruments, they finally made it to Dublin, later also Londonderry and Belfast. The Catholic Church then branded jazz as 'the music of the Devil' and further performances were banned throughout Ireland!

In December 1921 the Anglo-Irish Treaty, the solution for the Irish 'troubles', proposed the partitioning of Ireland and the creation of an Irish Free State. Unfortunately not everybody in Ireland was in favour of the proposal and civil war erupted in 1922. This had a damaging impact on trade on the Irish Sea. Passenger traffic had declined since the troubles first erupted in the mid-1900s, but now the cargo and livestock trades were also at risk.

On 25 July 1922 Coast Lines 'transferred' Laird Line Limited to G & J Burns Limited 'without price or other consideration'. Coast Lines Limited had acquired G & J Burns Limited in 1920. The Burns staff at Belfast were transferred to the Belfast Steamship Company Limited and that company then became Belfast agent for the newly merged company, which was given the title Burns & Laird Lines Limited (Chapter 15). The Laird funnel colours and the Burns houseflag, golden lion and globe on blue ground, were retained for the merged fleet. Routes and services were rationalised to generate efficient deployment of ships over a large area of the east, north and west coast Irish trade from the Clyde and elsewhere. Not only was this a logical merger of interests but it was also the best way forward for the new company to ride out the depressed trading conditions that now prevailed.

Analysis of all the accidents that occurred to ships of the Laird Line and its predecessors over the years clearly shows that the Firth of Clyde and River Clyde were the most dangerous areas frequented. There were 52 reported accidents (collisions, foundering, wrecks, etc.) in the River Clyde and 29 more in the Firth of Clyde. On the North Irish coast there were 32 accidents, while at the various Irish ports there were 26 at Londonderry, 19 at Sligo, 8 at Ballina, 9 at Belfast, 6 in the River Bann, 6 also at Westport, and 5 at Portrush. There were 15 accidents in the Irish Sea, a further 10 on the Lancashire coast, 5 in the Mersey and 4 more at Dublin and in Dublin Bay.

August 1921 - Laird Line 'Time and Fares Table'.

Olive (1893) was the largest ship in the Laird fleet when she was built. She remained on commercial service with her owners throughout the Great War.

[DP World]

Ayr Steam Shipping Company from *Clyde and Other Coastal Steamers* by Duckworth and Langmuir

Most of the early steamers of this fleet were called after rivers, and the names of nearly all had an intimate connection with Ayrshire. Mention may be made of the **Nith**, which, in October 1878, instituted a passenger service between Ayr and Larne and was wrecked on Ailsa Craig in November 1879; of **Lugar**, sunk by collision in September 1891, of **Doon** and **Afton**, acquired in 1877 (the latter foundered in October 1896); and of Laird's **Rose**, of 1867, later named **Ailsa**, which was wrecked on Portmuck, Islandmagee, on 26 February 1892. In March 1899, the Carron Company's ss **Margaret** was purchased and renamed **Mona**; she was wrecked in August 1908. Services were operated between Ayr and Campbeltown, calling at Kildonan in the south end of Arran. For example, in March 1886, **Rose** and **Lugar** were advertised from Ayr every Monday at 5.30 am, and from Campbeltown every Monday at noon.

Two three-masted ships, with machinery aft, **Garnock** and **Burnock**, were launched, respectively, in June and July 1890, and were owned by a subsidiary concern called Garnock Steamship Company Limited. These, together with **Doon**, were employed mainly on coastal runs with limestone cargoes. **Garnock** foundered on a voyage from Glasgow to Galway in December 1894, while **Burnock** was sold in 1901 to A Weir & Company, and in 1916 to the North of Scotland & Orkney and Shetland Steam Navigation Company who named her **Temaire**.

The Clyde Shipping Company's **Saltees** of 1885 was acquired in March 1892, and renamed **Carrick**, after the ancient name of the south western district of Ayrshire; and in the same year the Dundalk and Newry steamer **Amphion** was purchased. The former was sunk by collision with the Dublin and Glasgow company's **Duke of Gordon** on 25 May 1906, while the latter was sold in 1894 to P Mourout, Marseilles.

South Western, after leaving the fleet of the Ardrossan Shipping Company, was named in succession **La Valette** and **Quirang**. In 1897, she entered the Ayr Steam Shipping Company's fleet and received the name **Merrick** from one of the Galloway hills. She remained with her new owners about two years, being then sold to McDowall & Barbour, Piraeus, and renamed **Clio**.

CHAPTER 9

J & G BURNS; JAMES MARTIN & J & G BURNS; GEORGE BURNS – SHIP OWNERS

George Burns was born in December 1795. In 1818 he and his brother James began business as general merchants in Glasgow; by 1824 they had become joint owners, with Hugh Matthie of Liverpool, of six sailing vessels. Shortly afterwards, these sailing ships were replaced with steam ships powered with side-lever engines and driven by paddle wheels. In 1829, some of these ships were put into service in a highly competitive trade between Glasgow and Liverpool; others were employed on the route between Glasgow and Belfast. The latter trade was also highly competitive, but by 1830 the firm of J & G Burns had managed to secure a firm hold, by purchasing the interests of their principal rival John Gemmill. In 1834, however, the competitive struggle was renewed when ships owned by Messrs Thompson and MacConnell were put on the cross-channel service between Glasgow and Belfast. It was not until the middle 1840s that the Burns brothers began to secure a significant share of the trade and not until the 1850s that they were able to buy out Thompson and MacConnell, their rivals, and create a monopoly.

From: Francis Hyde, *Cunard and the North Atlantic 1840-1873*

In 1922 Burns was recognised as the dominant enterprise in the newly merged Burns & Laird Lines Limited, even though Burns was very much subordinate in terms of its longevity. But just as Alexander Laird and his son were pioneers in steam navigation, so too were the Burns brothers. Whereas the Lairds were content to work between Scotland and Ireland and England and Ireland, the Burns' interests crossed the Atlantic in company with Samuel Cunard and David MacIver. Nevertheless, there were numerous common interests that led to partnerships between the Burns and the Laird families, just as there were between Burns and M Langlands & Sons, and Burns and Thompson & McConnell. The collaborations thrived through both formal and informal Conference agreements so that the key players, at least on the Irish Sea, became a strong alliance ready to face any intruders.

The Belfast and Liverpool services to the Clyde were to become the core of the Burns' enterprise. However, both were to be highly competitive as trade developed, Belfast the more so, as it expanded rapidly from the 1840s when the Encumbered Estates Courts released the freehold on the Donegall and Chichester Estates. This action provided the land for extensive industrial development, including, of course, Donegall Quay, which became Burns' Belfast home.

The first steamship employed by Messrs J & G Burns was the ***Air***, of just 76 tons burthen. She was chartered from her owners and ran from Glasgow to Ayrshire and Galloway for the Burns brothers. Then, in 1824, they became agents for the Glasgow and Liverpool service operated by Matthie and Theakston. On 26 March 1826 the brothers sent the new steamship ***Fingal*** from Glasgow to Belfast; she was owned by a group of Glasgow businessmen under the title Belfast and Glasgow Steam Boat Company. James and George Burns were initially appointed as Glasgow agents but were soon also managing the steamer. The ***Fingal*** had sleeping berths for 30 passengers, could accommodate several horse boxes, which were lashed on deck, and carried up to 180 tons of cargo. The ***Fingal*** was scheduled to sail from Glasgow on Tuesdays and Fridays; the fare was 20/- in the cabin and 3/- on deck. Three years later, in March 1829, the Burns' brothers had created James Martin & J & G Burns which dispatched the steamship ***Glasgow*** from Glasgow to Liverpool.

The Burns brothers first became involved in the Liverpool trade in 1824. Until then three separate companies had run a fleet of sailing smacks between Glasgow and Liverpool. Six of the smacks were owned by Matthie & Theakstone of Liverpool, which acted as its own agent in Liverpool but employed John and Alexander Kidd at Glasgow. In 1824 Alexander Kidd died, and James & George Burns applied for the agency. Hugh Matthie accepted the Burns' offer. Shortly afterwards Theakstone elected to retire and J & G Burns bought his share to become equal partners with Matthie. In this way the Burns brothers had acquired the agency at Glasgow and had also become ship owners.

It did not take long for the brothers to persuade Hugh Matthie that steamships must be the way forward. They also approached Thomas and James Martin, who managed

George Burns 1795-1890.

James Burns 1789-1871.

a further six smacks in the Liverpool and Glasgow trade and they too were able to persuade their owners to join the venture to form James Martin & J & G Burns. On Friday, 13 March 1829, the newly-formed Glasgow and Liverpool Steam Shipping Company sent its first steamer, the **Glasgow**, on her maiden voyage from Greenock, followed a few weeks later by the **Ailsa Craig**, and in the following year by the **Liverpool**. The **Glasgow** and **Liverpool** were purpose-built, while the **Ailsa Craig** had been built in 1825 and was bought from Glasgow owners who had employed her on the Glasgow to Belfast crossing. In 1829 the service maintained departures from Greenock and Liverpool on Mondays, Wednesdays and Saturdays, but during much of the Autumn **Ailsa Craig** was not available following 'a slight accident', and a twice weekly service was offered.

In 1828 the new steamer **Eclipse** was chartered by J & G Burns for the Belfast route. She was needed to stand in for the **Fingal**, which was out of action following a boiler explosion while berthed alongside at Belfast. Two men were killed in the explosion. Shortly afterwards, the **Eclipse** was bought outright and registered under the ownership of the Belfast and Glasgow Steamboat Company and, when the **Fingal** returned, a two ship service was maintained, with departures from both ports on Mondays, Wednesdays and Fridays, normally at one o'clock in the afternoon. The Clyde terminal switched from the Broomielaw to Greenock when tides were unsuitable, and departures were put back to between 4 pm and 6.30 pm. A third ship was bought in 1830, when rival John Gemmill released his charter of the new steamer **Belfast**, allowing the Belfast & Glasgow Steamboat Company to acquire her. In so doing the main competitor was removed from the market.

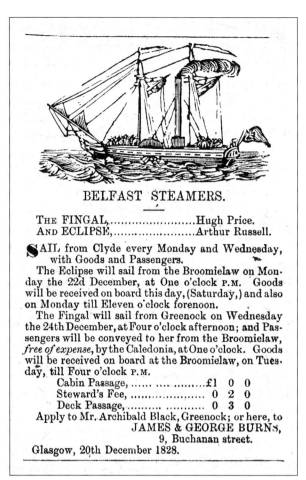

Advertisement for the Belfast service of J & G Burns with the steamers **Eclipse** *(1826) and* **Fingal** *(1819), 20 December 1828.*

The success of James Martin & J & G Burns, trading as the Glasgow and Liverpool Steam Shipping Company, is evident from their purchase of David MacIver's Mersey and Clyde Steam Navigation Company, the 'Manchester' company, and the main rival on the Liverpool service in 1831. With it came the steamers **Henry Bell, James Watt** and **William Huskisson**. These three new ships were initially used to maintain a daily service between Glasgow and Liverpool. However, it was soon realised that they were not up to the job and they were sold later in 1831 and 1832, the service reverting to three sailings per week. On 9 June 1831 the company bought the **Enterprize** from the New Clyde Shipping Company, thought better of the deal and sold her four days later. However, in 1832 two new purpose-built paddle steamers, the **Clyde** and **Manchester**, were delivered for the Liverpool service. The **Clyde** was distinctive as she was the first steamer to have the newly-developed steeple engine installed; a system with a vertical cylinder – a tall engine with a high metacentric height. Given the prototype engine, she was, by all accounts, a highly successful addition to the service.

Both the **Liverpool** and the **Glasgow** were chartered out in summer 1832 to run a service between Dundee and London. The arrival of steamers on the route prompted the local Dundee, Perth & London Shipping Company, which at that stage only operated a fleet of sailing smacks, to charter its own steamer and order two new ones, so the Burns' ships were soon returned to west coast duties.

Adverts in the *Liverpool Mercury*, 12 July 1833, included:

Steam packets for Greenock and Glasgow, with goods and passengers to the Broomielaw. The Glasgow and Liverpool Shipping Company's Steam Packets:
Liverpool   Robert Hepburn Commander
Manchester   H Main Commander
Clyde   Robt. Crawford Commander
Ailsa Craig   T Wylie Commander

and for 19 September 1834:

Steam Packets for Greenock and Glasgow. The Glasgow and Liverpool Shipping Company's fine Steam Packets
Manchester   H Main Commander
Clyde   Robt. Crawford Commander
Ailsa Craig   J Mckillar Commander
Glasgow   Thos. Wylie Commander
Liverpool   R Hepburn Commander

J & G Burns' Belfast and Glasgow Steam Boat Company was also doing well. The *Glasgow Herald* reported on 25 March 1833:

> We understand that Messrs Burns's splendid new steamer, the **Antelope**, for the Belfast trade, is to be launched today from Mr Barclay's Building Yard, Broomielaw, about half past three o'clock.

The **Antelope** (the first ship in the fleet with an animal name) was received from her builders in time for her maiden voyage to Belfast on 10 June 1833. Advertising in the Glasgow and Greenock press announced:

> This beautiful vessel is powered by two powerful engines, and is one of the swiftest sailers in this part of the country. Her cabin accommodation is of a very superior order, uniting at once elegance with comfort; and as it is intended she will sail from the Broomielaw twice or thrice a week, whilst the other vessels of the company will take intermediate days there will then be afforded an almost daily opportunity for the conveyance of goods and passengers between Glasgow and Belfast, and at freights and fares greatly reduced. Jas. & Geo. Burns, 9 Buchanan Street, Glasgow.

The arrival of the **Antelope** in service allowed the **Eclipse** to transfer to a new Greenock to Newry service in August. The larger **Belfast** replaced her in January 1834, when the St George Steam Packet Company started to call at Newry. However, the route was not a success and the service was closed a few months later.

Coincidently, another animal-named ship joined the James Martin and J & G Burns fleet in 1833, the **Gazelle**, purchased second-hand for the Liverpool route, but resold in the following year.

In 1835 another new steamer was commissioned for the Belfast route. This was the **Rapid**, which was registered under the new company title Belfast and Glasgow Steam Shipping Company. Other ships in the fleet were re-registered under the new company title over a period of several years. Arrival of the **Rapid** allowed the **Fingal** to be sold. **Fingal** had become the relief boat and had spent much of the previous three years out of steam and idle. The **Rapid** was sold in 1838.

In 1835 the **Eagle**, and in the following year the **Unicorn**, were commissioned for the Liverpool service. The **Unicorn** was typical of the day, had a carvel-built hull and was rigged as a three-masted schooner, later modified as a barque. She had one deck and poop, carried a unicorn head at the bow, and her square stern was decorated with mock quarters. Despite the lavish accommodation of the two ships, each reputedly costing £5,000 for the furnishings alone, they were sold in 1839 and 1840 respectively. The **Unicorn** was sold to the British & North American Steam Packet Company, in which George Burns had a considerable interest. She sailed across the Atlantic to run a branch service. Peter Newall in *Cunard Line – a Fleet History* described her voyage:

> In May 1840 the 649g paddle steamer **Unicorn** became the first ship owned by the British and North American Royal Mail Steam Packet Company. This tiny ship was bought for the mail and passenger feeder service between Pictou and Quebec. Pictou is about 100 miles to the north east of Halifax and the mails between the two towns were carried overland by stagecoach, a journey which took two days. Because of a delay in the completion of **Britannia**, it was decided that **Unicorn** should take the inaugural sailing for the new line. She moved from Liverpool's Clarence Dock into the River Mersey on Friday 15 May 1840 and the following day embarked the mails, newspapers and twenty-seven passengers, including Samuel Cunard's son Edward. After a rough crossing, **Unicorn** arrived at Halifax on Monday 1 June. During her brief stay in Halifax, around 3,000 people visited the ship before she sailed for Boston, arriving there on the 3rd. She berthed at the new 750 feet long Cunard Wharf, which had been erected by the East Boston Company. The wharf on Marginal Street was also close to the depot of the Eastern Railroad and the ferry to Boston city.

Short tenure in Burns' domestic fleets seemed to be the fashion, as the **Mercury** was bought second hand in 1836 and sold on in 1837, while the new steamer **Actæon** was delivered in 1837 and sold in 1841. She was bought by the Royal Mail Steam Packet Company for £23,599 but was wrecked three years later on a voyage from Santa Martha to Cartagena.

In 1835 the **Liverpool** was chartered for use on a service between Southampton and Guernsey. The ship gained quite a reputation with hard drinkers

The **Liverpool** (1830) seen in the colours of the Peninsular and Oriental Steam Navigation Company.

[From an oil painting by W J Huggins]

who had little respect for the finely appointed saloon. Respite came in August that year when she was sold and, in due course, deployed by the Peninsular Steam Navigation Company (later P & O) for their new service from Southampton to Alexandria. The *Manchester* and the *Clyde* were sold in 1837, the *Ailsa Craig* in 1838. In 1836 the *Manchester* was also chartered for the Iberian and Mediterranean services of the Peninsular Steam Navigation Company, carrying Prince Ferdinand of Saxe-Coburg to Lisbon for his marriage to Queen Maria II and, on return from Malta, brought home four giraffes for the Zoological Society of London. The *Manchester* remained on charter throughout 1837; she was returned to Burns in the following year and was sold to owners in Berwick.

A slightly smaller version of the *Actæon* followed in 1839, with the name *Aurora*. During trials, she suffered a list that seemingly had no easy remedy. On her maiden return voyage from Belfast on 30 January 1840 the thirty passengers demanded that the ship put back to Belfast and, off Carrickfergus, the master did so. The *Aurora* was off duty for several months while tests were carried out and suitable ballast installed to keep the vessel on an even keel. Another new steamer, the *Achilles*, joined the Liverpool fleet in mid-summer 1839. She was the largest and most powerful ship to date and represented the pinnacle in design of the wooden-hulled passenger paddle steamer.

Before the *Achilles* displaced the *Unicorn*, services were advertised as follows:

> *Actæon* and *Unicorn* (*Eagle* on standby) Monday, Wednesday and Friday departures from Glasgow and Liverpool – Cabin 20/-, steerage 7/-.
>
> *Aurora* – alternate Monday, Wednesday and Friday sailings from Glasgow and Belfast - Cabin 10/-, steerage 2/6d.

In 1842, J & G Burns bought the *Fire King* for use in the Liverpool trade. She dated from 1839 and had been used on coastal services, on which she had finally been overtaken by the new railways.

In September 1839 the boiler in the *Antelope* exploded while the ship was alongside at Glasgow. Fortunately there were no human casualties, but eighteen cattle were killed in the hold which was filled with explosive steam. Casualties were otherwise few and far between on both the Belfast and Liverpool services, although minor groundings and collisions were not uncommon. For example, the *Belfast* was reported aground on 2 March 1836 on the North Bank, off the Mersey, on a voyage from Liverpool to Wexford. She was under charter at the time and, although easily refloated, was found to have considerable damage to both hull and machinery. Later in the year, on 29 December 1836, the *Antelope* collided with the Drogheda steamship *Green Isle* in the Mersey, and went aground near Red Noses (near New Brighton), later to be refloated with considerable damage.

It is notable that every opportunity had been taken to upgrade both the Liverpool and Belfast ships, so that a modern and well equipped fleet would always be maintained in the two services. The reason why new and technically advanced ships were imperative for success was the intense competition on both routes. Thomson & MacConnell (David MacIver) and their City of Glasgow Steam Packet Company had the *Tartar* on the Belfast route, and their *Commodore* and *Admiral* on the Liverpool route. David MacIver's City of Glasgow Steam Packet Company, originally set up when the Manchester company failed, now worked alongside the Burns ships on the Liverpool service. This collaboration was typical of Burns, who preferred to work with, rather than against, competition.

M Langlands' Glasgow & Liverpool Royal Steam Packet Company opened a new service to Liverpool with *Princess Royal* and *Royal George* in March 1839. This provided additional competition on the Liverpool service and created an immediate rate war (see Robins and Tucker). The passenger fares were first reduced by half to saloon 10/-, second class with sleeping berth 5/- and deck class 2/6. Cargo in bales and boxes was carried for just 2d per cubic foot. By November 1839 they were halved again to 5/- saloon, 2/6 second class and 1/- deck class. Boxes and bales came down to 1d per cubic foot, and even a ton of cotton could be carried for as little as 10/-. Thomson & McConnell (David MacIver) and Burns both brought their fares down to match.

The new company also tried the Glasgow to Belfast trade with the *Royal Sovereign*, which completed five return voyages to Belfast in December, leaving Glasgow on Mondays and Fridays. Fares on this route tumbled to 5/- saloon, 3/6d second class and deck class 6d. The *Royal Sovereign* was advertised to maintain the service throughout January 1840, but in the event she stood down on Thursday 16 January.

By May 1840 the advertised fares to Liverpool were still depressed. Agreement was then reached that Burns and the new company (later to become M Langlands & Sons) would maintain sailings each way on a schedule that provided a daily service from both Glasgow and Liverpool other than on the Sabbath. The fare returned to 20/- saloon and 5/- deck class and boxes and bales went up to 2½d per cubic foot.

Two important initiatives took place in the 1830s quite divorced from the trades from the Clyde to Ireland and Liverpool. These were the involvement of George Burns in the West Highland trade from 1835 to 1851 and, in 1839, the co-partnership of George Burns with David MacIver and Samuel Cunard to create the British and North American Royal Mail Steam Navigation Company, later to become the Cunard Steamship Company.

George Burns entered the West Highland trade in 1835 when he acquired the **Inverness** and **Rob Roy**. William Young approached Burns to ask if he would be the Glasgow agent for these steamers which traded via the Crinan Canal and Caledonian Canal to Inverness. George Burns refused but countered with an offer to buy half shares in the ships with an option to buy the remaining half in the future should he so wish. The option was quickly taken up and, in September 1835, the transition of the business from William Young to George Burns and William Young and then to George Burns alone was completed in the space of just a few months. The early days of West Highland trade and the Burns' sixteen year tenure in the trade is summarised in *The Kingdom of MacBrayne*:

> The grouping of individual ships into small fleets becomes a recognisable trend after 1830. Robert Napier, a key figure in Clyde shipbuilding, was among the first to take an entrepreneurial interest in serving the West Highlands and the islands. In December 1835, the McEachern fleet became Napier's, along with the **Brenda** and part-ownership of the **Shandon**, two vessels closely associated with the Crinan Canal and linkages to Oban and Glasgow. In 1838, ownership passed to Thomson & MacConnell, who were agents for the City of Glasgow Steam Packet Company, founded in 1831. Following an agreement of partnership in 1840, this enlarged fleet merged with that of its competitors, James and George Burns…

As with the Liverpool and Belfast services, J & G Burns became the registered owner of the West Highland interests in 1842, rather than G Burns alone. *The Kingdom of MacBrayne* again:

> The merged West Highland fleet now included the Burns' steamers, the **Rob Roy**, the **Helen McGregor**, and the **Inverness**, all said to have been acquired in 1835. The Burns brothers soon took over other competitors, among them William Ainslie of Fort William, and in 1845 the Castles Company, latterly known as the Glasgow Castles Steam Packet Company, along with seven 'Castles', one 'Maid' and one 'Vale'. The Burns' Western Isles monopoly was now complete, and the brothers also retained a strong hold on the Clyde estuary routes.

> In 1846, a Burns vessel was placed at the disposal of the Admiralty to convey Grand Duke Constantine to view the islands. When it was announced that Queen Victoria was to make the same journey in the following year, another vessel was found and furnished from the Burns' household. The journey through the Crinan Canal (known to this day as part of MacBrayne's and Caledonian MacBrayne's 'Royal Route') was given a regal touch by the gaily decorated tracking-horses on the tow-path, with their attendants dressed in gold and scarlet.

> By 1851, the Burns' enterprise had become so unwieldy that they decided to concentrate on the Irish Sea and Atlantic services and to dispose of their Clyde and Western Isles interests. With the exception of five of the 'Castles', the fleet and its interests were transferred to David Hutcheson & Company of 14 Jamaica Street, Glasgow. The Hutcheson brothers, David (who had been employed by Burns since 1822) and Alexander (also a former employee of the Burns empire), were joined by George and James Burns' nephew David MacBrayne, who in 1851 had just turned 37 years of age.

In truth, the Burns brothers realised that the profitability of the West Highland trade was poor compared with the Clyde based Irish and Liverpool services. The transfer of the assets to David Hutcheson & Company, later to become David MacBrayne, was an act of faith in the Hutchesons and the young David MacBrayne, who was a nephew of George Burns. It was an act that reflected the burgeoning scale of trade on the Irish Sea and the rather static volumes of trade available on the Glasgow based routes to various West Highland destinations. A full account of all the ships operated under Burns' ownership in the West Highland trade is provided in the Fleet List, pages 120 to 124, and a full account of the Burns' West Highland involvement is given in Chapter 1 of Christian Duckworth and Graham Langmuir's *West Highland Steamers*. A full account of the subsequent Hutcheson and MacBrayne history is available in *The Kingdom of MacBrayne*, in which the financial collapse of David MacBrayne Limited in 1928 is described, along with the subsequent rescue of the company jointly by Coast Lines Limited and the London, Midland & Scottish Railway. It would seem that George Burns' vision of profitability back in 1851 was indeed a totally sound one.

The involvement of George Burns in the development of Samuel Cunard's vision to run regular transatlantic steam packets, unlike the West Highland initiative, was a most lucrative one indeed. The early days of the Cunard story are told in *Memoirs and Portraits of One Hundred Glasgow Men:*

> The Cunard Company is the most important undertaking the Burns's have had to do with. It dates from 1839. At this time the Admiralty had charge of our ocean mail service, and used to consign the mails for our North American colonies to the uncertain mercies of 'coffin brigs'. Steamers had repeatedly, had indeed twenty years before, crossed the Atlantic, but it was only in 1838 that the famous voyages of the **Sirius** from London and the **Great Western** from Bristol, decided the Admiralty in favour of steam. They issued circulars asking tenders for a steamer service to Halifax and Boston. One of these circulars came into the hands of a Halifax merchant, Samuel Cunard, who jumped into the next packet to see the Admiralty about it.

> Cunard was of an American family (originally Quakers), who having been Royalists at the time of the revolt of the Colonies, had taken refuge in Nova Scotia. He had a good position in Halifax: he was the agent to the East India

Co.: he was a gentleman: he was given to hospitality: a bright, tight little man, with keen eyes, firm lips, and happy manners. Halifax was then a great naval station, and Cunard had many friends in the fleet. This secured him a ready hearing at the Admiralty, and to the dismay of the Great Western company and the Bristol people, he carried off a contract for the mail service via Liverpool to Halifax and Boston. But he had not means himself to carry out the project, and he did not find much encouragement from London capitalists. Then he applied to Melvill, Secretary of the East India Company, whom he knew well through his Halifax agency; Melvill referred him to Robert Napier, who had built steamers for 'John Company'; Cunard came straight here to Napier; Napier, who had done work for our old friends the City of Glasgow Company, took him to its leading partner, James Donaldson; Donaldson took him to the Burns's; and the Burns's sent for David MacIver, and with him and Cunard went into the thing.

It was a big job for these days, and a risky one: the capital was to be £270,000 [£10,600 from James and George Burns]; there were to be steamers of not less than a stated tonnage and horse-power; to be set against the subsidy were heavy penalties if the passages were not made in a stated time; and the cost of driving full steam across the Atlantic in all winds and weathers was an unknown quantity. Finally, after a dinner with George Burns in Brandon Place, and a breakfast next morning with Napier at Lancefield, Burns and MacIver agreed to take it up if they could get a few friends to join them in a small private co-partnery, and could get the Admiralty to modify the contract in the way of an increase in the size and power of the steamers. Cunard agreed to everything, and George Burns set out on his rounds to find the partners. At the Lancefield breakfast he had asked for a month to fill the list. He had underestimated his hold on the public confidence. His first call was on William Connal, who gave him a short answer – 'I know nothing of steamers, but as you say it is a good thing, put me down'. Others followed suit, and the list was made up with ease within four days. Then John Park Fleming sat up all night, and with his own hand wrote out the contract of co-partnery, and we do not suppose any of the co-partners had any reason to regret having put their hand to it.

The Admiralty contract, as finally adjusted, was in the names of Samuel Cunard, George Burns, and David MacIver. It was for two sailings per month from Liverpool to Halifax and Boston, and the subsidy was £60,000. These terms were afterwards largely increased on both sides, but from the first the Cunard Company (or, in its formal title, the British and North American Royal Mail Steam Navigation Company) gave extra measure. The first vessel ready was the **Britannia**, and she sailed from Liverpool on her first voyage, on Friday, 4th July, 1840, Independence Day as it happened. Cunard sailed in her; he was uneasy over the Friday sailing; but in spite of it the **Britannia** arrived safe, and from that day to this the Cunard Company have never lost a life nor a letter.

First of the Cunard paddle steamers, **Britannia** (1840), on her maiden departure from Liverpool on 4 July 1840 with 64 passengers, arriving Halifax, Nova Scotia, twelve days later.

The subsequent story of the Cunard Line, wholly a story of success, is legend and well documented. New services to the Mediterranean were developed from 1853, largely through David MacIver's existing interest in that area. The company later saw off the competition from the American Collins Line, which collapsed in 1858.

The original investments made by the co-partnery were amply rewarded. The diverse ownership of the British and North American Royal Mail Steam Navigation Company was soon bought out and eventually came down to just three families. Mediterranean services were developed and, when these were merged with the North Atlantic interests In 1866, ownership of what shortly became the Cunard Steamship Company was vested in Charles MacIver, George Burns' two sons, John and James Clelland Burns, and Sir Edward Cunard and his brother William. The Burns sons each held about £175,000 worth of shares, together owning one third of the company. In 1879 these investors sold their interests in the British and American Royal Mail Steam Packet Company at the incorporation of the Cunard Steamship Company Limited; the Burns brothers each received 200 £1,000 shares in the new company. G & J Burns was then appointed agent for the Cunard Line at Glasgow, MacIver at Liverpool and William Cunard at London. Shares only became available to the general public in 1880, when a prospectus was issued.

Nevertheless, riches accrued from the Cunard enterprise did not deter the Burns family from pursuing its Clyde-based interests in trade to Ireland and Liverpool. Indeed the family did this more and more earnestly as time went on.

CHAPTER 10

G & J BURNS' DIVERSE TRADE INTERESTS

One famous line the Burns's worked up, and then, under the pressure of greater enterprises, abandoned. This was the Royal Route (well-named) through the West Highlands. They began the Highland trade about 1832, and in 1835 bought three little steamers, the **Rob Roy**, the **Helen Macgregor,** and the **Inverness**, which William Young, a plumber, had been running to little profit through the Crinan. From this beginning they worked up a whole system of steamers for the day passage through the Crinan or the night passage round the Mull, gliding along canals or battling with the Atlantic, meeting at Oban, crossing and re-crossing, plunging into the lochs, winding along the sounds, threading their way among the islands, fine pleasure boats for the flock of summer swallows, stout trading boats summer and winter serving the whole archipelago, linking with the world the lonely bay or the outer islet, freighted out with supplies of all sorts and shapes, freighted in with wool and sheep, Highland beasts and Highland bodies: surely the liveliest service in the world! But they had their hands more than full, and in 1851 they handed over the whole fleet to the new firm of David Hutcheson & Co.

From Memoirs and Portraits of One Hundred Glasgow Men

In January 1842 the Liverpool and Belfast interests, along with the West Highland services were merged under the title G & J Burns, George having clearly taken the initiative with the shipping side of the business while James had stayed partly tied to the original victualling business. The Liverpool agent was Thomas Martin and Burns & Company, James Martin having retired. The Liverpool and Belfast fleets then comprised: **Antelope**, **Aurora**, **Achilles** and **Fire King**, such that in January 1842 the **Fire King** maintained the Belfast service under Captain McKellar, and the **Achilles** maintained a twice weekly Liverpool service under Captain Main, and the two older ships were out of service during the winter months. For much of 1842 Thomson & McConnell had the **Commodore** or the **Admiral** on their Liverpool service and Langlands had **Royal Sovereign** and **Royal George**.

The joint service with Thomson and McConnell via Crinan, left both the Broomielaw and Inverness every Wednesday morning at 7 am. This route, along with services from Oban, was maintained by the **Shandon**, Captain McLean; **Brandon**, Captain MacKillop; **Helen McGregor,** Captain Turner; and **Rob Roy**, Captain Duncan. The summer season was more intense with **Helen McGregor** and **Rob Roy** leaving Glasgow for Inverness every Monday and Thursday at 5 am, the **Brenda** running from Oban to Loch Etive with trips also to Staffa and Iona, and the **Shandon** on the Glasgow to Dunoon and Kilmun route on the Clyde.

In 1843 the **Aurora** was on the Belfast service, still offering a cabin fare of 10/- plus 2/- steward's fee. On the Liverpool service was the **Achilles**, while Thomson & McConnell had just the **Admiral** on the Liverpool service and Langlands had commissioned the new paddle steamer **Princess Royal** to run alongside the **Royal George.**

A seemingly odd second hand purchase in 1844 was the acquisition of the elderly steamer **Foyle** from the Glasgow & Londonderry Steam Packet Company (see Chapter 2).

The first iron-hulled steamer to be ordered by Burns was the **Dolphin**. She was designed for the West Highland services and was an impressive looking steamer complete with three masts. In 1845, her first season, she worked out of Oban. In October she moved to the Glasgow to Belfast route to inaugurate a daylight service, leaving Glasgow at 9.30 am on Mondays Wednesdays and Fridays, returning on alternate days at the same time from Belfast. It was an unfortunate time of year to trial such a venture, as passenger numbers were low; the service was stopped after just one week. In the spring the **Dolphin** returned to Oban. Meanwhile the **Aurora** maintained the overnight Belfast sailings while **Antelope** and **Achilles** were surplus to requirements and were sold, the latter to the Peninsular and Oriental Steam Navigation Company.

Having tried out the iron-hull concept on the Belfast daylight service, a new steamer was ordered specifically for the service. This was the **Thetis**, another three-masted and schooner rigged ship with a neat profile such that the mizzen mast was well aft. Her hull was slightly narrower than normal such that her length to breadth ratio was 8.6 compared with the traditional design of 8 or less. She was specially equipped with a powerful side lever engine, which was intended to maintain a speed of 16 knots. She took her maiden voyage from Greenock on 4 August 1845, at the height of the passenger season. During the week she made two daylight return crossings from Greenock and two from Glasgow, returning to the Broomielaw overnight. She only carried passengers who were charged 10/- cabin and 2/- steerage. Again, however, the speed of the ship was just not quite a match for the demands the schedule placed on her and she was withdrawn in October and placed on the normal overnight service.

Burns now had only the **Fire King** and **Aurora**, on the Belfast and Liverpool routes respectively and the **Thetis** working on either service as required. Burns still had its expansive fleet of West Highland steamers, having taken over the Clyde interests of the Glasgow Castle Steam Packet Company in 1845, that company's ships being integrated into the

Burns' fleet three years later. This acquisition gave Burns the all-important link between Glasgow and Ardrishaig at the end of the Crinan Canal. In 1847 Burns also acquired the Bute Steam Packet Company and its three ships **Pioneer**, **Pilot** and **Petrel**.

Burns was conscious that Langlands had the prestige steamer **Princess Royal** on the Liverpool service and that the company needed a match for it. This was the **Orion**, famous not because of her design or economic success but because of her tragic ending in 1850. She was launched on 19 December 1846 from the yard of Caird & Company at Greenock. **Aurora** then stood down from the Liverpool station and was replaced by the brand new paddle steamer **Viceroy**, chartered from the Dublin & Glasgow Sailing & Steam Packet Company. **Aurora** then displaced the **Fire King** from Belfast duties; she was sold for further use. **Aurora** worked the Belfast service on alternate nights with **Thetis**. The **Foyle** was sold shortly afterwards.

In due course the **Orion** was fitted out and ready for her maiden voyage from Greenock to Liverpool on 12 May 1847. The **Viceroy** had been replaced on charter by the **City of Dogheda**, dating from 1836, and the charter was ended as soon as **Orion** was ready for service. The second ship on the service was Thomson & McConnell's **Commodore**. She had a far more traditional hull than the earlier **Thetis** with a length to breadth ratio of 7.8. She was powerful. She was also well-appointed, and she quickly became popular with the travelling public. On 23 October 1848 the **Orion**, inwards to Liverpool from Glasgow, ran down and sank the sailing smack **Riviere**, which was on passage from Hayle in Cornwall to Runcorn.

Orion (1846) was wrecked off the Galloway coast with great loss of life in June 1850.

[Illustrated London News]

A new paddle steamer, the **Lyra**, was commissioned for the Belfast service in 1848. She displaced **Aurora** from the service so that **Aurora** became spare ship, although she was little used. A second new paddle steamer, the **Camilla**, joined the **Lyra** in 1849; she was bought from the builders, Caird & Company, who had also built the **Lyra**, at auction for just £11,500. Commissioning of the second new ship allowed the **Thetis** to be chartered to Thomson & McConnell to run alongside the Burns' steamers on the Liverpool route, although she also worked some services to Belfast. **Aurora** was retained, although little used.

In July 1849 G & J Burns obtained the contract for carrying the mail between Scotland and Northern Ireland. Until then the mail had been carried by Government-owned steamers operating between Donaghadee and Portpatrick, the shortest possible link but one requiring extensive use of stage coaches. The new mail contract specified a daily overnight service Monday through Saturday, and Burns agreed that the mails would be carried free of charge (a fee chargeable by weight was payable by the Post Office only from 1882 onwards, replaced by an annual fee in 1883). This rather magnanimous offer was made for fear that the Ardrossan Steam Navigation Company might win the contract with its two steamers **Fire-Fly** and **Glow-Worm**. Until 1849 three sailings a week were given by Burns and a fourth by Thomson & McConnell's **Tartar**. With the mail contract in place, the steamers **Camilla** and **Lyra** operated three return trips a week.

Minor accidents continued to occur; on the morning of 4 September 1849 the **Camilla** on passage from the Clyde to Belfast, went ashore near Larne, but was refloated on the flood and arrived at Belfast without apparent damage.

Lloyd's List 2 January 1850, reported:

> 28 January 1850, Glasgow: The **Camilla** (s), for Belfast, in going down the river, when off Govan, was in collision with a Steamer having in tow the **Cora Linn Lambert**, from New York, and all the vessels received considerable damage.

In 1849 G & J Burns acquired the business of William Ainslie of Fort William and his three paddle steamers **Curlew**, **Queen of Beauty**, which Burns renamed **Merlin**, and **Maid of Morven**. *The Queen of Beauty*, when originally built, was an engineering curiosity, as James Williamson described in *Clyde Passenger Steamers 1812-1901*:

> The **Queen of Beauty**, built by Thomas Wingate in 1844, and engined by Robert Napier, was the idea and property of John Kibble, the original owner of the Conservatory known as the Kibble Palace, in Glasgow Botanic Gardens. The boat was fitted with two paddle-shafts on each side, one forward and one aft, between two and three feet above the water line. On each shaft a drum was fixed, and a belt with floats fastened to it extended from one drum to the other. This was intended as an improvement on the ordinary paddle-wheel, but it did not prove a success. She was afterwards fitted with the less ambitious paddles…

Bell of the *Orion* (1842).
[McLean Museum, Greenock]

The last addition to the Burns West Highland fleet was the **Lochfine**, acquired in 1850 from William Roxburgh. She was notable as the first screw steamer in Burns' large and diverse fleet, and the first screw steamer in the West Highland trade, for which she had been built in 1847. She was built by William Denny & Brothers for William Roxburgh (Glasgow & Lochfyne Steam Packet Company) and regularly traversed the Crinan Canal, for which her screw propulsion and absence of sponsons was a great advantage. She did have a bowsprit, being schooner rigged, but did not carry a figurehead which could have obstructed lock gates.

John Kennedy recorded the loss of the company flagship **Orion** in *The History of Steam Navigation*:

> One of the most convincing proofs of the splendid management of the several steamship companies which trade between Liverpool and Glasgow, of the skill and honest workmanship put into the vessels, and of the great care exercised by the officers who navigate these ships, is the fact that for upwards of eighty years there has been but one disaster accompanied by loss of life on this station.
>
> The disaster referred to was described at the time as 'a disaster occasioned by the most culpable carelessness'. It was, unfortunately, attended with frightful loss of life. The **Orion**, a splendidly fitted and powerful steamer, sailed from Liverpool for Glasgow, on Monday afternoon, 18th June 1850, with about 170 passengers in addition to a crew of 40 all told. It was an ideal summer trip; the night was fine and clear, and the sea perfectly smooth. All went well with the steamer and those on board until, about a quarter past one on the Tuesday morning, the sleeping passengers were rudely awakened by the concussion, as the ship struck violently on the rocks, close to the Lighthouse at the entrance to Portpatrick Harbour. The vessel, which was steaming at full speed at the time, filled instantly, and sank in a few minutes. The night was so tranquil that many of the passengers had slept on deck, but the majority were asleep in the cabins below when the catastrophe occurred. The scene of horror and dismay which followed can be but faintly conceived. A wild rush of crew and passengers was made to the boats. The first boat lowered to the water was so crowded instantly with panic-stricken passengers, that she capsized, and all who were in her were drowned. A second boat was launched, in which some ladies were placed, and these reached the harbour safely. One redeeming feature in this tragic narrative is the splendid heroism displayed by many of the gentlemen passengers. The second boat when launched was in the first instance filled by men, but when the officers of the ship suggested to them that their first duty was to save the women and children, most of the men instantly left the boat, and assisted females to occupy the places they had surrendered, who were thus happily preserved.
>
> Shortly after this boat got away the ill-fated **Orion** sank, and all on board either went down with her, or were left floating on the surface of the water, or clinging to floating portions of the wreck.
>
> The Ardrossan and Fleetwood steamship **Fenella** passed the scene immediately after the disaster occurred, and the Captain at once stopped his ship, lowered his boats, and rendered valuable assistance in saving lives. The

Lighthouse keepers and Coastguards had also observed the vessel coming too close in shore, and, anticipating a catastrophe, had awakened the local boatmen. Owing to this, numerous boats had instantly put off, and these picked up a large number of those floating. By the continued efforts of the **Fenella's** crew, and the Portpatrick boatmen, about 150 persons were rescued. This dreadful catastrophe carried mourning into many of the most respectable families in Liverpool and Glasgow. Amongst those who perished were Captain McNeil (brother of the Lord Advocate), his wife and two daughters; Dr. Burns, one of the most popular men in Glasgow, professor of Surgery at the University, and brother to the Managing owners in Glasgow; Miss Morris, his niece; and Master Martin, a son of one of the Liverpool owners.

The trial of the Captain, and first and second mates of the **Orion**, for the 'culpable bereavement of the lives of the passengers' who were lost by the wreck of that steamer, as before narrated, took place at Edinburgh, before the High Court of Justice, on the 29th August 1850. It was proved that during the second mate's watch, the vessel approached closer to the shore more than was usual by upwards of a mile, and that this course was maintained notwithstanding the warning exclamations of the experienced seamen who were on the look out. It was further proved that the Captain had come on deck several times during the second mate's watch, and each time had observed both the compass, and the ship's proximity to the shore, which could be clearly seen, and yet did not countermand the second mate's instructions. The charge against the first mate was withdrawn, but at the end of the trial, which lasted two days, the Court sentenced the Captain to be imprisoned for eighteen months, and the second mate to be transported for seven years.

Her master was Captain Thomas Henderson who, after serving his gaol sentence, later went on to become a key promoter of the Anchor Line. He never forgave George Burns for his lack of support and he promoted the Anchor Line in direct competition with the Cunard Line both on the Atlantic and the Mediterranean, even naming his ships with the suffix –ia as did the Cunard company.

Lloyd's List reported on 12 December 1850:

10 December, Greenock. The **Thetis** (s), from Belfast, was in contact in the morning of 8 December off Pladda with the **Lavinia**, hence to Santa Cruz and Matanzas, and received considerable damage in her bow. The damage sustained by the **Lavinia** was above the water line, and it is understood she will bear up for Rothesay Bay or the Pool for repairs. The **Thetis** proceeded and ran ashore in dense fog a little below Kilcriggan Quay, on the north bank of the Firth, but got off on the flood and reached Dalmuir.

Shortly afterwards, early in 1851, the whole of the Burns network of Clyde, West Highland, Belfast and Liverpool services were completely overhauled. Christian Duckworth and Graham Langmuir described the changes:

At the end of the previous year Messrs Thomson & McConnell had retired from the Liverpool and Belfast trades, which then passed into the hands of Messrs G & J Burns, Messrs Burns & McIver being agents for that company at Liverpool. The West Highland services, which had been operated jointly by Messrs Thomson & McConnell and Messrs Burns since 1839, were transferred in February 1851 to Messrs David Hutcheson & Company; and Messrs Burns retired from the Clyde river trade. Certain of their Clyde steamers were retained by their purchasers, Messrs Wm. Denny & Bros., the Dumbarton shipbuilder. These were **Culloden**, **Petrel**, **Dunoon Castle**, **Rothesay Castle**, **Inverary Castle**, **Cardiff Castle**, **Craignish Castle**, **Dunrobin Castle** and **Merlin**; and some of them soon found other owners, mostly on the Clyde. **Pilot** also, was sold; and Messrs Hutcheson received the West Highland steamers **Dolphin**, **Pioneer**, **Cygnet**, **Lapwing**, **Duntroon Castle** and **Curlew**, along with the Crinan Canal trackboats **Sunbeam** and **Maid of Perth**.

The Burns fleet was, accordingly, very much depleted, and consisted (after the above mentioned disposals) of **Camilla** only. **Commodore** was chartered [in 1850] to take the place of the wrecked **Orion** on the Liverpool station; and the Dublin steamer **Æriel** was chartered in 1851 until the new steamer **Stork** should be ready, **Camilla** being then transferred to the Belfast run. **Thetis** came back to the Burns fleet from Thomson & McConnell.

So just what had precipitated such a massive reorganisation and depletion of company assets? It was certainly not shortage of cash; the new ship for the Liverpool station, the iron-hulled paddle steamer **Stork**, was delivered in May 1851, albeit smaller than the **Orion** which she replaced, but equally as well appointed and equipped.

The critical factor was the retirement of Thomson & McConnell. This left Burns, at last, with a monopoly on the Glasgow to Belfast service, in competition, although in concert, with Langlands, on the Liverpool station and pretty much alone on the West Highland routes where the Thomson & McConnell ships had also passed to Burns. Burns had also obtained the mail contract for the Belfast station, albeit at no remuneration to G & J Burns, but a prestigious milestone for that daily service. There was growing competition on the Clyde services with the railways and railway company-sponsored shipowners, such that Burns probably considered himself well out of that trade, hence his withdrawal from all but the Ardrishaig service which went to Hutcheson. Besides, the Clyde and West Highland services had really got out of hand, starting in 1835 with just three ships with an aggregate tonnage of just 62 tons and within 15 years it had increased to 24 ships with an aggregate tonnage of 2,462. Indeed the only reason Burns had purchased the Glasgow Castle Steam Packet Company in 1845 was to acquire the all-important link between Glasgow and Ardrishaig, with connecting steamers via Crinan to Oban. It is really no surprise that Burns later wanted to dispose of the other services, principally from Glasgow to Bute, by selling the fleet in 1851.

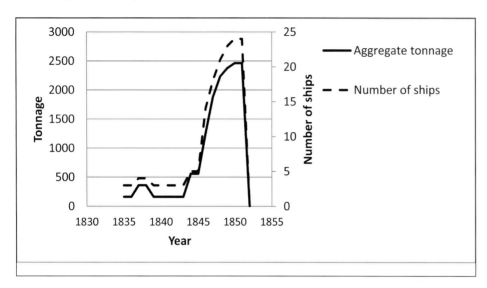

Figure 1: Development of Burns Clyde and West Highland interests between 1835 and 1851.

The impetus for the reorganization in 1851 was the favourable prospects of the monopoly on the Belfast Mail Service, coupled with the powerful Burns/Langlands collaborative service on the Liverpool run. Demand for these two routes, both passenger and cargo, was increasing rapidly and their economic potential was considered to outweigh anything that the West Highland services could offer.

David Hutcheson & Company was formed in February 1851 to receive ownership and management of the former Burns ships and company goodwill. David and Alexander Hutcheson and David MacBrayne were the partners of the new company, the Hutcheson brothers being former managers in G & J Burns and MacBrayne a nephew of the Burns brothers. Precisely what financial caveats George Burns placed over the new partnership is unclear, but from the outset, at the beginning of February 1851, the new company appeared to be financially independent from G & J Burns. David Hutcheson's obituary published in 1880 in *The Engineer* shows insight on the relationship with the use of the word 'invited':

> In the year 1824 Messrs G & J Burns went into the Liverpool steamship trade, and gave Mr. Hutcheson an important position in carrying it on. By-and-by the same firm established an important connexion with the carrying trade of the West Highlands, including Tobermory, Skye, Staffa, Inverness, etc. But their Liverpool trade, and the trade which they had also established with certain Irish ports, and more especially their great Atlantic traffic in the Cunard Line, became so immense that they deemed themselves justified in retiring from the West Highland steamship enterprise about thirty years ago. The success of the boats running to the West Highlands was comparatively small until the year 1851, when Mr. Hutcheson was invited to undertake the management of the concern, which he did in partnership with his brother, Mr. Alexander Hutcheson, and Mr. David MacBrayne.

As a 'gift' to the new partnership it was at best a magnanimous act of a successful businessman, at worst it was a two edged sword. If Hutcheson, as a senior manager in the firm, had been expecting a partnership prior to the big sell-off, then perhaps he was given a bargain offer, that he could not refuse, as compensation for what he had anticipated would be his but was being denied. However, the new company was grossly undercapitalized and in desperate need of new, efficient and up-to-date tonnage and considerable investment was needed. It was probably this reason that led George Burns to retire not only from the Clyde routes but also from the West Highland routes. Indeed George Burns was looking further afield, not only across the Atlantic to North America, but also to the Mediterranean in collaboration with David McIver and latterly Cunard. The depletion of the Burns network also meant that management could concentrate on two core products, Clyde to Belfast and Clyde to Liverpool, without any dilution of effort.

Before leaving this curious divestment of assets, it is interesting that David Hutcheson & Company was able to find funds to replace the ships on the prestige Glasgow to Oban service. The **Mountaineer** was specially built for the Clyde section of the route in 1852 and, three years later, the first of the three famous Clyde steamers with the sacred name **Iona** displaced even the crack ship **Mountaineer** to other duties. There were even funds available to build a new steamer for the Glasgow to Western Isles via Mull of Kintyre service – the new paddle **Chevalier**. George Burns may have been reluctant to invest in such tonnage, but others were not so reserved, as funding is not reported as an issue, perhaps largely because at that time business was expanding on the Ardrishaig service and good returns were to be had by investors. This was a reflection of the burgeoning emigration trade from the western Highlands as a consequence of the Clearances, whereby sheep essentially replaced crofters, and of the increased summer tourist traffic. The new company, later restyled David MacBrayne, enjoyed the rest of the nineteenth century and only started to run into financial trouble after the Great War, when the then flagship **Chieftain** had to be disposed of in a fire sale to support the ailing company. The business never fully recovered and it was reconstructed in 1928 and brought partly within the Coast Lines group.

G & J Burns' next priority was consolidation of the daily Belfast Mail Service. David McNeil wrote:

> When the Burns monopoly began in the early 1850s, the **Thetis**, **Lyra**, **Camilla** and **Stork** were regularly employed in the mail service. Each weekday one of the vessels sailed from Belfast at 7 pm with passengers and mail for the Clyde. Contemporary newspaper advertisements continually emphasized that this departure would be punctually observed and warned passengers not to be late. On the return journey the steamers left the Broomielaw at Glasgow at 4 pm and, after a call at Greenock, proceeded direct to Belfast. The Greenock call, a feature of the Glasgow route until the 1914-18 war, enabled passengers to go between Glasgow and Greenock by rail and therefore save several hours on the journey between Belfast and Glasgow.

Further consolidation occurred when three new sister ships were built for the service in 1853 and 1854 by William Denny & Bros at Dumbarton and named **Elk**, **Stag** and **Lynx**. The *Illustrated London News* reported of the **Stag**:

> This fine vessel is one of three magnificent new steam-vessels, respectively named the **Stag**, **Elk** and **Lynx**, now fitting out for the Glasgow and Belfast Royal Mail Steam Packet services. The mails between Scotland and Ireland have been conveyed by this route since July, 1849, with a regularity and convenience unknown before, when the mails between the two countries were carried in Government steam-vessels between Port Patrick and Donaghadee.
>
> The service was undertaken by Messrs Burns, of Glasgow, and has been carried out by them free of charge to Government by their steam-vessels employed for the conveyance of goods and passengers between these important districts of the two countries – thus conferring an important benefit on the community by their regularity and efficiency, and at the same time saving a large sum of money to the country every year by the withdrawal of the expensive and inefficient packet stations formerly maintained.

Stag (1853) built for the Glasgow to Belfast service and sold to become a Confederate blockade runner in 1864.

[Illustrated London News]

The three new paddle steamers each had clipper shaped bows, three masts and twin funnels placed abaft huge paddle boxes. They were equipped with side lever engines and paddle wheels with fixed floats; these were the last Burns steamers to be equipped in this way. The saloon was an exceptional 65 feet long and was decorated in the heavy panelling and drapes preferred in that era. There were berths available in staterooms for 70 saloon passengers. The **Camilla** and **Lyra** were then sold, while the **Thetis** had already been disposed of during 1851 as part of the great 'sell off'. The three new sisters, **Elk**, **Stag** and **Lynx**, introduced for the first time the concept of identical ships that were interchangeable on a given service. The concept looked forward to replacing the group when they had outrun their usefulness. It would appear that George Burns was responsible for introducing this logical idea that became universally accepted once shipowners had acquired the resources to invest in blocks of new ships.

In the meantime, the Liverpool service was maintained by the **Stork** in conjunction with Messrs Langlands' service. G & J Burns had become an established company focused on its Clyde to Liverpool interests and the Clyde to Belfast mail service. It had shed itself of the cumbersome network of Clyde and West Highland services and various other diverse interests, including the Loch Lomond Steam Boat Company and the Caledonian & Dumbartonshire Junction Railway, that it had collected during the 1840s, and had emerged a leaner and better targeted company. As such it was well placed to face the downturn in trade that followed in the mid-1850s.

CHAPTER 11

CONSOLIDATION

The first purpose-built iron-hulled screw steamer in the fleet of G & J Burns was the **Beaver**. She was a small, compact cargo steamer intended to supplement peak sailings on either the Liverpool or Belfast services. She was commissioned towards the end of 1854 and, in addition to supplementing existing services, was used to inaugurate a new cargo service between Belfast and Preston. The wharf on the Ribble at Preston had few facilities and was poorly connected to serve the industrial development then taking place in Lancashire. The service was an ambitious challenge which competed head on with the railway company services into Lancashire, and Burns soon withdrew from Preston. The **Beaver** was acquired by Burns and MacIver for their new Liverpool to Havre service. She was resold to Peter Denny as shipbroker in autumn 1856.

Although the **Beaver** was not a commercial success, her screw propulsion clearly found favour with the owners, who then ordered two more screw steamers to serve the Liverpool trade. These were the **Otter** and **Zebra**, which were both commissioned in summer 1850. They were not sister ships, although both were built at Greenock, the **Otter** by Robert Steele & Company and the **Zebra** by Caird & Company. The **Otter** was sold on completion, due to the downturn in trade that occurred on the route in the early 1850s, but the **Zebra** was commissioned for service and did operate, albeit briefly, between the Clyde and Liverpool. Towards the end of 1855 the **Zebra** transferred to the MacIver service to Havre, working alongside the **Beaver**.

The **Zebra** was lost at Carnvoose Cove on the Lizard peninsula in July 1856. *Lloyd's List* reported:

> Plymouth 23 July. The **Zebra** (s), from Havre to Liverpool, struck on the rocks near the **Lizard** [22 July], filled and sank immediately; crew and passengers saved.

> Falmouth 23 July. Some baggage, and about 100 packages of cargo, and some ship's stores, from the **Zebra** (s.s.), [Captain] Beth, from Havre to Liverpool, which struck on some rocks near the Lizard, yesterday, have been saved; exertions are being made to save the ship and the remainder of the cargo.

> Falmouth 1 August. The **Beaver** (s), from Liverpool, arrived here yesterday, to take on board the recovered cargo of the **Zebra**, and to render assistance in endeavouring to get the latter off the rocks; a diver with needful apparatus has also come round with the **Beaver**.

> Falmouth 17 August. The **Zebra** (s), ashore at the Lizard, broke in two about the forward engine room yesterday, owing to the strong east wind and the heavy ground swell.

A new ship, the **Panther**, was commissioned on the Liverpool service in 1857. She was a reversion to paddle propulsion and her only modern feature was the lack of bowsprit and figurehead, encumbrances that only got in the way when docking at Liverpool. She was of essentially similar design to her running mate, the second **Princess Royal**, that had been completed for Langlands, also in 1857. It was not long before the new paddle steamer was chartered to the Peninsular & Oriental Steam Navigation Company for service in the Mediterranean. She was later purchased by the Italian Navy. Her place at Liverpool was taken by yet another paddle steamer, the **Leopard**. She was similar again to the **Panther** and **Princess Royal** and offered little by way of innovation.

Another screw cargo ship, with steerage accommodation only, the **Harrier**, joined the fleet in 1858. She was bought in a near complete state from her builders, William Denny & Brothers at Dumbarton. The **Harrier** was used variously on a revived Belfast and Preston cargo service and she also supported the Liverpool service during the summer months. Most importantly the **Harrier** was able to carry on the Liverpool service alone from June 1859 when the **Leopard** was transferred to the Belfast to Clyde route at the height of the season in June (see loss of the **Elk**, below).

A number of minor accidents occurred to the three Belfast sisters, their masters desperate to maintain schedule whatever the weather. The **Elk** ran aground in the Clyde on 13 March 1855. Before she was refloated, the steamer **City of Baltimore** struck her near the bow, removing the bowsprit and causing other significant damage. The **Stag** went ashore at Blackhead, at the entrance to Belfast Lough, on 9 March 1857, but she too was refloated and was soon back in service. *Lloyd's List* 10 March 1857 reported:

> Belfast 9 March. The **Stag**, of and from Glasgow, for this port, with a general cargo, ran ashore on the rocks, near Blackhead, during a snow storm, yesterday morning; the **Cambria**, from Ardrossan, went to her assistance, and took off the passengers; the **Stag** then backed off, and with assistance from the **Cambria**, reached the quay, with the food compartment full of water.

On 23 April 1857 the **Micmac**, which was approaching Belfast, was run into by the **Lynx**; there was no loss of life. On 19 February 1858 the **Lynx** arrived in the Clyde from Belfast without her mizenmast and with her rudder damaged, having been in contact with a large ship (supposedly American), off Pladda. One week later, on 26 February, the **Lynx** was in collision with the sailing ship **Oriental**, which was approaching Greenock from London. Neither ship was seriously damaged, although the **Oriental** lost her martingale. On 22 February 1858, the **Lynx** ran foul of the **Candia** off the Irish coast, causing substantial damage to the forepart of that ship, which had to be towed into Belfast by a tug. Finally the **Stag** touched ground in Belfast Lough, as reported in *Lloyd's List* 24 March 1858:

> Belfast 22 March. The **Stag**, Stuart, of and from Glasgow, for this port, with a general cargo, in coming up the Lough yesterday morning, during a thick fog, struck on the Horse Rock, Greypoint, came off on the following tide, apparently not damaged and sails tonight with mails, etc.

Worse was to come on 7 June 1859, when the paddle steamer **Elk** took the ground at high tide near Groomsport, on the southern shore of Belfast Lough. The accident happened at about 4 am as the vessel approached Belfast in a thick fog. *The Shipwreck Index of the British Isles* concludes:

> **Elk**, voyage Greenock – Belfast, 54.40.45N, 05.35.30W. Sailed from Greenock at 20.05 on 6 June, carrying passengers, mail and a general cargo, passing Pladda lighthouse at 23.00. At 01.30 she met with thick fog, and an hour later orders were given to allow the boiler fire to burn down, which reduced steam pressure and the vessel slowed down, but no further precautions were taken, nor was the lead and line used. Shortly afterwards she struck the rocks at Ballymacormick Point, and became a total loss. Her stranding and wreck was attributed to the default of the Master in failing to use the lead and line, and for not slowing the speed of the vessel whilst in fog. As a consequence, Captain MacQueen's certificate was suspended for twelve months at the BOT enquiry held at Glasgow. The passengers, crew and cargo were landed safely, the wreck being sold at auction to a Mr Madden, of Belfast, for £1,020.

The Liverpool to Glasgow paddle steamer **Leopard** was quickly drafted onto the Belfast mail service to replace the **Stag**. The 'cargo steamer' **Harrier**, which had been partnering the **Leopard** for the summer season on the Liverpool service, and advertised as 'for steerage passengers only', then became the only Burns ship on the Liverpool route for the next eight months.

In 1860 George Burns, who was then 65 years old, commenced divesting himself of his many and diverse managerial roles, which were taken over by his eldest son John Burns, who was then 31. The 1860s became a decade characterised by John Burns ordering numerous new vessels, few of which lasted in the fleet for very long. Some, of course, sailed westwards to become blockade runners for the Confederates during the American Civil War.

The first pair of ships were remarkable in that they were sister ships and because they were also the first screw steamers designed for the main Clyde to Liverpool service. These were the identical sisters **Heron** and **Ostrich**, Yard Numbers 73 and 74 at William Denny & Brothers, Dumbarton. They arrived on station in February and March 1860; advertisements in the press advised 'these new steamers have splendid accommodation for cabin and steerage passengers'. Although the new screw steamers were slower than the paddle steamers, they could still manage the passage from Greenock in little over 16 hours. They were also far more fuel efficient than the paddlers and their machinery required a great deal less maintenance; paddle floats required attention after almost every voyage was completed. Initially, until the **Ostrich** was ready for service under Captain Davies, the **Heron**, under Captain Brice, worked alongside the **Harrier** on a revived two ship, three sailings per week service, part of an integrated schedule maintained with Langlands. The **Harrier** was sold in May.

Heron *(1860) or* **Ostrich** *(1860) in Tod & McGregor's dry dock at Meadowside in 1862.*

The *Glasgow Herald* reported on 17 May 1860:

> Launch of the **Giraffe**: The launch of this steamer, the future of which excites so much interest by reason of well-grounded expectations that she will realise a higher degree of speed than has yet been attained in British or American waters, took place yesterday morning from the building yard of Messrs. J & G Thomson, Clydebank, Govan in the presence of a select party of ladies and gentlemen…
>
> She will be propelled by a pair of oscillating engines of 300 horse-power collective, with feathering paddles and will be supplied with six tubular round boilers… The steamer on nearing the water received the name of **Giraffe** from Mrs Margaret Blackburn, of Glasgow College [an accomplished author]… The **Giraffe** will accommodate with ease and comfort about 200 Cabin and 600 Steerage Class passengers. It is nothing new to state that Messrs Burns' line of Royal Mail steamers between Scotland and Ireland is one of the best organised and most efficiently conducted services in the world. The ships are manned and equipped with the utmost care and liberality; and the cleanliness and discipline on board are not exceeded by the similar state of things in His Majesty's line of battle-ships. We may expect, therefore, that the **Giraffe** will be fitted out and handled in a style every way calculated to secure speed, safety and comfort.

Shortly afterwards, the new and much needed fast paddle steamer **Giraffe** was commissioned for the Belfast service, under the command of Captain Stuart. She was designed with a service speed of 18 knots to revive the daylight service from Greenock to Belfast, but was also for use on the overnight mail service. The **Giraffe** carried out a trial trip on 10 July, leaving Greenock at 8 am; the *Glasgow Herald* reported the next day:

> The **Giraffe** made a great run in five hours and forty five minutes against a head wind. She attained the greatest speed between the Cumbraes and Pladda, when she was going grandly, but after which one of the engines heated and retarded her. She has no vibration and is remarkably steady at sea… The **Giraffe** met a hearty welcome at Belfast from the Harbour Commissioners and other Gentlemen, who came on board… The **Giraffe** started on her return voyage at about 4 o'clock, and arrived at Greenock, after a stoppage at Wemyss Bay, and a detour to Dunoon, before 11 o'clock.

Thereafter, the **Giraffe** was advertised to leave Glasgow every Saturday at 6 am, and Greenock on the arrival of the 7 am train from Glasgow, for Belfast. She was scheduled to leave Belfast at 2.30; the fare of 12/6d was the same as that for the overnight mail steamer, but steerage passengers paid a premium fare of 4/- against the overnight steerage fare of 3/-. During weekdays the **Giraffe** was rostered onto the mail service to support the **Lynx**, **Stag** and **Leopard**. Despite all the fanfare, the new ship struggled to maintain her seasonal day return service and it was not resumed in 1861.

Giraffe (1860) was designed for a daily return trip between Greenock and Belfast; in reality she never quite maintained the speed to do this.

[Illustrated London News]

The **Giraffe** was not ideally suited to the overnight service, as she did not have enough sleeping berths to satisfy normal demand. Although she never managed her expected 18 knots after her initial trial run, she did attract the attention of the agents of the Confederate States, who bought her in 1862. Renamed **Robert E Lee**, she sailed westwards with a full cargo of military stores. On 9 November 1863, the USS **James Adger** captured the blockade-runner **Robert E Lee** off Cape Lookout Shoals, North Carolina. The **Robert E Lee** had left Bermuda on 7 November with a cargo of shoes, blankets, rifles, saltpeter and lead. Her captain, Lieutenant John Wilkinson, CSN, later wrote: 'She had run the blockade twenty-one times while under my command, and had carried abroad between six thousand and seven thousand bales of cotton, worth at that time about two millions of dollars in gold, and had carried into the Confederacy equally valuable cargoes.'

Between July and August 1861, the **Leopard** was chartered to the Belfast Steamship Company and made ten round trips between Belfast and Liverpool. She was replacing the Belfast company's **Semaphore**.

The next steamer to be delivered for the Belfast Mail Service was the **Wolf**, a handsome, single funnel paddle steamer with a simple two cylinder engine. The *Glasgow Herald* reported on 22 November 1862:

> This beautiful vessel was launched from the building-yard of Messrs Napier yesterday afternoon. She is the first of Messrs Burns new fleet of steamers now building for the mail line between Glasgow and Belfast, is larger and more powerful than the **Giraffe**, and will be a worthy successor to that celebrated vessel.

The **Wolf** allowed the **Leopard** to be released for blockade-running bringing valuable dollars into the Burns' bank account. The **Leopard** was renamed **Stonewall Jackson**. On 11 April 1863 she attempted to get into Charleston and dashed past USS **Flag** and USS **Huron**, both of which fired after her with some shells penetrating the hull. With no escape possible from the port the **Stonewall Jackson** was run aground and destroyed with her cargo, including Army artillery and some 40,000 pairs of shoes. All personnel and the mails were landed safely.

Two identical paddle steamers were ordered from Caird & Company at Greenock to replace the **Stag** and **Lynx** on the Belfast Mail Service. They were equipped with oscillating engines which would allow them enough speed to maintain the service schedule with ease. Allocated the names **Roe** and **Fox**, neither ship ever sailed for Burns to Belfast. The **Roe** was launched as **City of Petersburg**,

Giraffe (1860) under conversion to a blockade runner in Tod & McGregor's dry dock at Meadowside in 1862.

The *Penguin* (1864) served G & J Burns for 14 years before she was sold to the Union Steamship Company Limited of New Zealand as seen in this picture.

while **Fox** retained her name when both ships headed west to the Confederate States. The **City of Petersburg** eventually found her way home to become Laird's **Garland** (see Chapter 5).

Lucrative vessels they may have been, but the Belfast Mail Service was in real need of upgrading. Repeat orders were immediately placed with Caird & Company and a new **Roe** and **Fox** were finally ready to work alongside the **Wolf** in 1864. However, after the **Roe** had been commissioned and the **Stag** had stood down ready for sale, it was decided to keep the **Lynx** in service and sell the new steamer **Fox** back to her builders, presumably at some profit. Caird & Company then reconfigured the ship as the blockade-runner **Agnes E Frey**. The service was then maintained by the **Wolf**, **Lynx** and **Roe**.

While all this reshuffling had taken place on the Belfast service, the **Heron** and **Ostrich** had maintained the two-ship service between the Clyde and Liverpool without upset. In March 1864 the **Heron** was sold and was replaced by the **Blenheim**, on charter from the Belfast Steamship Company. In April, the new passenger cargo screw steamer **Penguin** arrived on station under the command of Captain Bryce. A new consort for the **Penguin**, the smaller steamer **Beagle**, was commissioned by Captain Dugald McPherson with a maiden voyage from the Clyde on Wednesday 2 November 1864, returning from Liverpool on 5 November. Her arrival allowed a three ship service to be carried on until the **Ostrich** stood down for maintenance in the New Year. On her return, the **Beagle** transferred to Belfast cattle runs, although she was back on the Liverpool run in the summer.

Unhappily the new **Beagle** was run down and sank, just over a year after her maiden voyage. She had left the Liverpool service in the hands of the **Penguin** at the end of the summer season and had been on the Belfast to Glasgow service with extra cattle runs. Her schedule was 12.00 noon departure from Glasgow Mondays and Thursday (not calling at Greenock) and returning from Belfast at 10.00 pm on Tuesdays and Fridays. The *Glasgow Herald*, 10 November 1865, reported:

> In the '*Herald* of yesterday, an intimation, which had been received by telegraph, was made of a collision having occurred in the channel on the previous night, between the screw steamer **Beagle**, belonging to Messrs Burns, and the screw steamer **Napoli**, owned by Messrs Handyside & Henderson. From information since obtained, we gather that the **Napoli** was at the time bearing down channel, for Malta, Trieste and Venice, while the **Beagle** was coming up from Belfast for Glasgow. The night was clear, when at some distance the **Napoli** was observed bearing down upon the other vessel. About 7 o'clock, while off the Cumbraes, the two steamers came into collision, the shock being such that the **Beagle** shortly afterwards sank in deep water, and the **Napoli** was very seriously damaged. The **Beagle** was employed by the Messrs Burns as an extra steamer for carrying goods between Belfast and Glasgow, and the pleasant fact that no lives were lost may probably be attributed to the circumstances that no passengers

The *Camel* (1866) was one of three similar paddle steamers delivered to Burns in 1865 for the Londonderry, Belfast and Liverpool services.

[Contemporary painting of unknown provenance, Mike Walker collection]

were aboard at the time. The crew of the **Beagle** were picked up by the steam tug **Pearl**, which fortunately happened to be passing inshore of the other vessel, and were landed at Greenock. The damages sustained by the **Napoli** were of so serious a nature that it was found necessary to bring her back to Glasgow for repairs.

The **Beagle** was built only last year, and that she gave satisfaction to the owners is evidenced by the somewhat singular coincidence that only on Wednesday last the Messrs Burns instructed Messrs James & George Thomson to build a screw-steamer similar to the vessel which was fated to be lost on the evening of the same day.

On 1 March 1866 J & G Thomson launched the ship ordered in November 1865 with the name **Weasel**. She was indeed similar to the **Beagle**, having almost identical dimensions and was ready for service in June.

Three new paddle steamers were commissioned from Caird & Company, at Greenock, to maintain the overnight mail service to Belfast. These were the **Buffalo**, **Llama** and **Camel**, near sisters, and each equipped with oscillating engines to sustain a design speed of 13½ knots. The **Buffalo** took her first sailing to Belfast in August 1865, the **Llama** joining her later in the year, allowing the **Roe** to be sold. The third paddle steamer, the **Camel**, was not ready until the following summer. The **Buffalo** was taken off the Belfast route in March 1866 to inaugurate the new Burns service from Glasgow to Londonderry (Chapter 5) and then in June she moved to the Liverpool route, while the **Weasel** took over Londonderry duties.

In mid-January 1866, the **Ostrich** was sold. Her place on the Liverpool service was taken by McConnell & Laird's screw steamer **Falcon** from 17 January and, from the second half of February, by the **Warsaw**, chartered from Donald Currie of Leith. The **Warsaw** was replaced by the **Weasel** in the last week of April and the **Warsaw** was returned to her owners. **Weasel** and **Penguin** now partnered Langlands' screw steamers **Princess Alice** and **Princess Royal**. The Clyde Shipping Company's **Tuskar** replaced the **Weasel** in mid-May, in order to provide increased passenger accommodation on the service, and finally the Belfast paddle steamer **Buffalo** came on service alongside the **Penguin** for June and July. In August a new screw steamer, the **Snipe**, was delivered for the Liverpool station and the parade of different steamers finally ended. The **Snipe** took her maiden voyage from Glasgow on Monday 13 August, under the command of senior master Captain Dugald MacPherson, and returned from Liverpool two days later. She was well balanced alongside the **Penguin**, both steamers offering saloon and steerage accommodation, along with a large deadweight capacity. Ironically, Langlands had need to charter in the spring when the steamer **Liverpool** was on the joint service. The upset caused by the sale of the **Ostrich** in January does beg the question why she was released at that time; the answer can only have been an extremely high price being offered.

The following year the mail steamer **Wolf** sank after a collision in Belfast Lough. The *Glasgow Herald* 17 October 1867, reported:

> The **Wolf** left Belfast shortly after 9 o'clock on Tuesday night, with, so far as could be judged, about 100 steerage and from 25 to 30 cabin passengers. A dense fog prevailed in the lough, and for fully three hours the vessel crept along very slowly. A quarter of an hour or so after midnight and, when nearly opposite Carrickfergus, Capt Macaulay gave orders to cast anchor. These orders were immediately obeyed, and until after five o'clock in the morning the vessel remained stationary. Not long after the hour mentioned she was again got under way, and was proceeding slowly along when the captain observed the Belfast and Fleetwood steamer **Prince Arthur** coming up the lough. He ordered instant backing of the engines and the captain of the **Prince Arthur** was heard to cry 'hard a port'. The orders were given too late, however, to prevent a collision. The **Prince Arthur** struck the **Wolf** on the starboard bow, some ten feet aft; cutting right in, and sticking to her. Captain Macaulay, having shouted to the master of the **Prince Arthur** not to attempt to back, the **Wolf**'s passengers got on board the Fleetwood steamer. An impression exists among some of those who arrived in Glasgow last night that two or three adults were drowned. Some of our informants speak of two infants having perished…
>
> Meanwhile, the boats had been lowered, and the crews of both vessels exerted themselves to secure luggage which was tossing about on the water, and to save the cattle which were swimming about in all directions. Ropes were thrown around some of the animals, and they were swung on deck. After the lapse of a little time, the steamer **Countess of Eglinton,** from Ardrossan, came alongside and all the passengers – both those who had taken out passages on the **Wolf**, and those that had come from Fleetwood on the **Prince Arthur** – were transferred aboard her. She reached Belfast in due course. The **Prince Arthur** parted from the **Wolf**

Wolf (1863) being raised in Belfast Lough where she had sunk following a collision with the Fleetwood steamer *Prince Arthur* in 1867.

[*Illustrated London News*]

about twenty minutes after the collision, and the latter immediately sank. Though considerably damaged, the former vessel was able to make her way to Belfast. Our informants add that the only portions of the **Wolf** visible in the lough yesterday were the tops of the masts.

Amazingly, the hapless **Wolf** was eventually recovered from where she lay. She was refurbished and put back in service as the mainstay on the Londonderry service.

The last two ships to be built for G & J Burns in the 1860s were both ordered from J & G Thomson of Govan. They were the paddle steamer **Racoon**, which was designed for the Belfast service, and the screw steamer **Raven** for the Liverpool service. The **Racoon**, like her predecessors that had also been built by Caird & Company, was equipped with a twin cylinder oscillating engine, powerful enough to attain 17 knots on trials. She was commissioned in November 1868, allowing the **Lynx** to be disposed of shortly afterwards.

David McNeil wrote:

> She had a gross tonnage of 800 tons and her length, 252 feet, was not exceeded by any Burns steamer until the arrival of the **Adder** in 1890… She was well proportioned with two funnels, one fore and one aft of the paddle-boxes; these were large but, nevertheless, blended in with the lines of her hull to give her a distinctly handsome appearance.

The **Raven** commenced her maiden voyage to Liverpool on 12 January 1869, allowing the **Weasel** to be disposed of shortly afterwards. The **Raven** was a big passenger and cargo steamer that operated on the Liverpool to Glasgow route almost continuously for the next twelve years.

In September 1869, the **Snipe** was chartered to David Hutcheson & Company to replace their steamer **Clansman**, which had been wrecked in July. The **Snipe** later returned to her owners, who deployed her on the Belfast service until she was sold in January 1870 for further service between India and the Persian Gulf, there being a new screw steamer on order from J & G Thomson at Govan (Chapter 12). The **Snipe** left the Clyde on 7 February under Captain Pinhey, called at Cairnryan and left there for Lisbon on 10 February. She had a rough passage to Lisbon, losing stanchions and other deck fittings. Some of the cargo in the forward hold had to be removed to effect repairs. She was reported to have arrived at Port Said on 8 March and by 22 March she had arrived at Bombay and was handed over to her new owners.

Thus, at the start of the 1870s the fleet stood at seven ships. The screw steamers **Penguin** and **Raven** were on the Liverpool station, and the paddle steamers **Wolf**, **Buffalo**, **Llama**, **Camel** and **Racoon** formed the core of the Belfast mail service. The spare ship from either service was used for the Londonderry to Clyde route, and the screw steamer **Penguin** was also frequently deployed on Belfast duties, particularly when demand for cargo and cattle shipments was high. The fleet was modern and up to date, certainly fit for purpose, and by all accounts popular with travellers, not least for its perceived safety record.

July 1870 advertisement for G & J Burns daily Glasgow and Greenock to Belfast service with onward train connections, .

CHAPTER 12

SCREW STEAMERS TO BELFAST

J & G Thomson, at Govan, delivered three new screw steamers to Burns in 1871 and 1872. The first, the **Bear**, was launched on 12 September 1870, the **Bison** on 20 April 1871 and the **Ferret** on 24 November the same year. The **Bear** was the first ship in the fleet to be equipped with a compound engine. She was used primarily on the Clyde to Liverpool route, but was also deployed on the Clyde to Londonderry service during the peak summer season. Thus in summer 1871 she was rostered to leave Glasgow at 4 pm every Wednesday and Saturday, returning from Londonderry on Monday and Thursday. The saloon fare was 12/6 and steerage 3/-. The **Bison** was a slightly larger version of the **Bear** and she commenced on the Liverpool service alongside the **Raven** on Friday 4 July. Her arrival on station allowed the **Penguin** to stand down and take up duties on the Belfast Mail Service. The arrival of the **Penguin** at Belfast allowed the **Wolf** to be withdrawn and sold. Finally the **Ferret** was designed for the summer only Belfast daylight passenger and cargo service, run as a supplement to the overnight mail boats. She also deputised for the paddle steamers on the mail service off season. The **Ferret** left Greenock's Albert Harbour on arrival of the 11.15 am train from Glasgow every Monday, Wednesday and Friday, returning from Belfast at 8.00 am every Tuesday and Thursday and at 12.00 noon on Saturday.

It was all change on 21 November 1871, the year of James Burns' death. The Belfast paddle steamer **Buffalo** replaced the **Bear** on the Londonderry service and the **Bear** replaced the **Bison** to work alongside the **Raven** on the Liverpool route. The **Bison** had completed just seven months in her owners' service, working the Liverpool overnight route before she was sold. Again, a favourable offer for the almost new ship must have tempted the Burns' directors and, in January 1872, she sailed for duty on the Caribbean feeder routes of the Royal Mail Steam Packet Company. Unlike many of her predecessors, the next new ship, the **Owl**, remained in service for a long 22 year period, almost all of it on the Liverpool route. She was a big passenger and cargo ship, equipped with a compound engine. She was managed jointly by G & J Burns and David McIver, an illustration of the degree of collaboration these two companies now maintained. The **Owl** took her maiden voyage from Glasgow on Monday 6 October 1873, working alongside **Raven**. Her arrival on station relieved the **Bear**, which transferred to the Londonderry service. The **Penguin** was variously used on Belfast sailings and as relief for the Liverpool steamers. The paddle steamer **Buffalo** was still working the Clyde to Londonderry route, and the other paddle steamers, plus the **Ferret**, were used on the overnight mail service to Belfast. The **Ferret** was released at the end of October and sold to the Dingwall & Skye Railway without change of name. She attained notoriety in 1880 when she was stolen from her then owners, the Highland Railway, the thieves being apprehended in Melbourne, Australia, some months after the ship had disappeared.

The **Ferret** (1871) seen in the colours of the Adelaide Steamship Company which she served for nearly 40 years.

[Thompson State Library, South Australia]

Bear (1870) was used principally on the Glasgow to Liverpool service but also served between Glasgow and Londonderry.

[DP World]

In 1874 a new, but small, cargo ship, with limited saloon and steerage accommodation, was commissioned as a replacement for the **Ferret**. She was given the name **Hornet** at her launch, on 4 March 1874, from the yard of Blackwood & Gordon, at Port Glasgow. An almost identical ship followed her down the same slipway later that year and was christened **Wasp**. The pair were built as a response to the increased cargo traffic on the Belfast service, and had been designed for an overnight supplementary service for cargo and cattle, which would take the pressure off the mail ships and their tight schedule. The **Hornet**, however, started on the Londonderry service, allowing the **Buffalo** to return to Belfast duties. In October the **Bear** took over the Londonderry roster and the **Hornet** and the **Wasp** started on their new overnight cargo runs between Glasgow and Belfast – without a call at Greenock. A third cargo steamer, the **Jackal**, was launched in August 1875, but was never commissioned by Burns; she was completed as the **Taiaroa** for service in New Zealand. The **Hornet** was sold in 1876, while the **Wasp** was put under the joint management of G & J Burns and M Langlands & Sons to run a highly successful cargo-only service between Glasgow and Liverpool, later becoming wholly owned by Langlands until, in October 1879, her ownership passed to John Langlands, Mathew Langlands, John Burns, Glasgow, Henry Lamont, Glasgow, & Francis Johnston, Liverpool. She was wrecked in the Mersey in July 1888 (see Robins and Tucker).

Poor weather continued to cause disruption and there were numerous minor collisions, groundings and other incidents. One of the more alarming reports was that in *Lloyd's List* 23 February 1877:

> Greenock. The **Bear**, which arrived here, yesterday, from Londonderry, with cattle, experienced very heavy weather after passing Innishowen Head. Heavy seas struck the vessel on the port side, washing clean over the deck, and the supports on which the lifeboats sit beside the davits were washed away from the boats on the port side. On arrival of the steamer at Greenock about 40 head of cattle, it is stated, were dead.

By 1877, the demand on the Belfast service was such that two ships were required each night to clear all the passengers, cargo and cattle needing to cross the channel. From mid-May 1877, a second steamer was put on the overnight Belfast Mail Service. One of the two sailings called for train passengers, as before, at Greenock, while the other sailing went direct to Ardrossan to pick up train passengers, the advantage being a 10 pm connecting train departure from Glasgow rather than the traditional 9.05 pm train to Greenock. The Ardrossan connection proved troublesome, with the existing operator, the Ardrossan Shipping Company, aided by local interests, doing everything they could to delay the Burns steamer. The call was dropped in favour of Greenock two weeks later, and so began the Burns year-round, two ship departure service from Glasgow and Greenock to Belfast, the second ship offering a later train connection from Glasgow to Greenock.

The **Penguin** was the first screw steamer to be employed regularly on the Belfast Mail Service for any length of time, a role she adopted when she was displaced from the Liverpool route in July 1871. Her success in working alongside the faster paddle steamers and, more particularly in terms of passenger comfort in heavy seas, finally convinced her owners that screw steamers were viable on the route. An order was placed for two sisters, to be named **Walrus** and **Mastiff**. The pair were near sisters of the **Owl**, but had slightly less depth, so that they could steam up Belfast Lough at full speed, even at low tide. They were needed to support the double sailing introduced in May 1877 from both Belfast and the Clyde, with the second service leaving on weekday evenings only, either at the same time as the mail ship or shortly after it.

The two state-of-the-art passenger and cargo screw steamers acted almost in a relief capacity. It is interesting to note that, at that time, technology could not allow a ship with sufficient deadweight capacity to be designed that would be capable of replacing the two-ship service, and yet operate economically within the navigational confines of Belfast and the Clyde approaches. The **Walrus** commenced service in May 1878, and was followed by her sister in July; the faithful **Penguin** was then sold. The **Walrus** did not last long in service to Burns and was sold in 1880 to the Greek Navy, in anticipation of a cluster of new ships then being built for the Belfast route.

On 22 June 1878, John Burns and his wife, accompanied by sixteen guests, including writer Anthony Trollope, used the new steamer **Mastiff** for a sixteen day cruise to Iceland. The party returned to Wemyss Bay in the afternoon of Monday 8 July and, after strawberries and cream at Castle Wemyss, which had then become the Burns family home, they departed. Trollope later described the trip in his book *How the 'Mastiffs' went to Iceland*, for example:

> Captain Ritchie, was a nautical authority great in ice, and peculiarly conversant with the Northern seas… we had a crew of thirty two men, including engineer, seamen, stewards, and firemen; and thus, being fifty on board, we started from just beneath the towers of Castle Wemyss on the evening of Saturday, the 22nd June, 1878. I presume it will be known to all who may read these pages that Castle Wemyss is the abode of our host, Mr John Burns.

> …the **Mastiff** and all that was in it was the property of Mr Burns, and that we were his guests from the moment we put our foot upon the deck till we left the vessel, after the lapse of three weeks, again in Wemyss Bay, under the walls of his residence. We had been summoned for Saturday, the 22nd, and were all on board the vessel that day at six pm. A few minutes later the anchor was weighed, amidst the mingled cheers and lamentations of our friends on the shore, lamenting, not that we were going, but that they should not have been able to accompany us.

The trip enthused John Burns to order a steam yacht, the **Matador**, of 220 tons gross, and she was the first of eight yachts owned within the family over the next four decades. The Burns family were, by all accounts, now fabulously rich, but such riches did not deter John Burns or his sons from putting all their energies into the management of both G & J Burns and the Cunard enterprise.

Mastiff (1878) served Londonderry for much of her career before migrating to the Manchester to Glasgow route in 1895. She is seen on the left at Pomona Docks, Manchester.

The Burns directors were always on the lookout for opportunities to expand their services. One such came about when a group of Larne traders proved the viability of a passenger and cargo service between Larne and Glasgow. MacNeill wrote:

> [The service] began in January 1877 when the locally-owned *Larne* began a thrice-weekly service between the two ports. She generally sailed from Larne at 4 pm and the fares charged were 7s 6d saloon and 3s deck. In the autumn of the same year the sailings were reduced to two per week. In August 1878 J & G Burns placed their small steamer *Rook* on the same run and six months later the *Larne* was withdrawn. The *Rook* ran twice a week and carried both cargo and passengers. After a short time the service became cargo only.

The *Rook* was launched as *Moorfowl* in early June 1878, but was completed with the name *Rook*. She was built specifically to run in competition with the *Larne* and, as such, had berths for a small number of saloon passengers, as well as steerage accommodation. The bully boy tactics worked and the Burns house flag became a regular twice weekly sight at Larne. The *Rook* sailed on Mondays and Thursdays, from Glasgow at 4 pm and Greenock at 7 pm, returning from Larne on Tuesdays and Fridays at 8 pm. The saloon fare was 12/6 and steerage was 3/-. The *Rook* remained on the Larne station until spring 1880, then transferred to Liverpool duties for a short while before she was sold in January 1881. Her place on the Larne station was then taken by any spare ship from the Belfast route.

The paddle steamer *Llama* went ashore in March 1881 and, although later refloated, was so badly damaged that she was sold for demolition. The subsequent Board of Trade Enquiry, found neither the Master nor First Mate to blame for the incident and reported:

> The *Llama* was an iron screw steamer, built at Greenock, in the county of Renfrew, in the year 1865. She was registered at the Port of Glasgow, of 686.78 gross, and 391.32 registered tonnage.
>
> She was owned by Mr John Burns, of Jamaica Street, Glasgow, and was under the command of Mr Dugald McPherson, who holds a certificate of competency, No. 100,010.
>
> The *Llama* left Belfast at 8.35 pm on 3rd of March bound for Glasgow, manned by a crew of 31 hands, all told, a small general cargo of about 10 tons, and no passengers. Her draught of water was stated to be 9.9 [feet] aft, and 8.6 forward. At 10 pm the same evening the patent log was put over off Blackhead, and the tide being then flood, and the wind from SE threatening, a course was shaped ENE, allowing for both wind and tide, to make Corsewall Light, the master very prudently wishing to get into shelter under the land, and so await for finer weather to proceed on his voyage.
>
> The wind increased, after leaving the Lough of Belfast, to a severe gale, accompanied by blinding snow storms. The most vigilant look-out for the land was kept, and the master, conceiving that he had run his distance, was in the act of telegraphing to the engine room to stop the engines, when high land was suddenly seen right ahead, and simultaneously with the order to stop the vessel stranded on the bold coast of Wigtonshire.

The master ordered rockets to be fired, but these signals failed to bring assistance, and at daylight two messengers were dispatched for the same purpose, but they also failed, the snow being too deep to enable them at that time to proceed. Subsequently the little cargo there was on board was sent to Glasgow, but the sea and the weather remained too boisterous to make any attempt to get the vessel off. Eventually she was floated and taken in safely to Stranraer.

The end of the traditional Belfast paddle steamers finally came later in 1881, when three fine and fast screw steamers were delivered for the mail service. The three near identical sisters were built by Barclay, Curle & Company at Whiteinch and given the names **Alligator**, **Dromedary** and **Gorilla**. They were bigger than the **Walrus** and **Mastiff** and, with more powerful engines, could manage a service speed of 14 knots on the overnight Belfast run. These ships had excellent saloon passenger accommodation, with a large dining saloon, still arranged around the traditional long table. The *Glasgow Herald*, 13 January 1881, reported on the launch of the **Alligator**:

> This vessel, which is intended for Messrs Burns service between Glasgow, Greenock and Belfast, was launched yesterday from the building yard of Messrs Barclay, Curle & Co., at Whiteinch. On the signal being given, the dog-shores were knocked away, and the ship glided swiftly into the water, the ceremony of naming her being performed by Miss Muir, daughter of Messrs Burns' Superintending Engineer.

Gorilla (1881) and *Hare* (1886), moored to the right, above the York Street ferry.

> The **Alligator**, as a cross-channel steam ship, is of large size, her dimensions being – length, 280 feet; breadth 30 feet; depth 15½ feet; and she will be fitted by her builders with engines of great power, which combined with the beauty of her lines and model will ensure for her an exceptionally high rate of speed. Besides possessing extensive cargo capacity and a spacious steerage, as well as cattle dealers sleeping cabin, she has a very large poop saloon which is to be handsomely fitted with berth and dining accommodation for upwards of 100 first class passengers.
>
> The **Alligator** is the first of three sister steamers now in course of construction at Messrs Barclay Curle & Co. for Messrs Burns. These vessels have been specifically designed to meet, by every modern appliance of modern architecture and adaptation of science, the requirements of the Scotch and Irish mail service, and, along with other two steamers now building for Messrs Burns by Messrs Blackwood & Gordon at Port Glasgow, will be completed and ready for use in a few months, when they will all be placed on the station between Glasgow, Greenock and north of Ireland ports. The unusual speed of these new vessels will so diminish the time of the passage, that Messrs Burns will be enabled, not only to sail from Greenock and Belfast at a later hour than at present, and still preserve all their train connections inland for passengers, mails and goods, but to deliver cargo in Glasgow and in Belfast even earlier than by the existing arrangements.

The steamers building at Blackwood and Gordon's yard were the smaller *Lizard* and *Locust*. These sister ships were designed primarily for supplementary cargo runs to the overnight mail service, but they also offered limited saloon and steerage passenger accommodation. The *Lizard* was launched on 16 April 1881 and the *Locust* exactly a fortnight later. By the end of May the Belfast and Londonderry services were in the hands of screw steamers at long last, with the *Alligator*, *Dromedary*, *Gorilla*, *Mastiff*, *Locust* and *Lizard* in charge. Another cargo ship, the *Lamprey*, a modified version of the *Lizard* and *Locust*, was building at Blackwood & Gordon's yard, and she joined the service later in the year.

This massive investment in new ships allowed the disposal of the three remaining paddle steamers, most of which still had several years' useful service ahead of them. The **Buffalo** and **Camel** went to Barclay Curle as part payment for the new passenger and cargo steamers, and with new engines installed were sold to the Barrow Steam Navigation Company as the **Donegal** and **Londonderry**. The **Racoon** had also gone to Barclay Curle the previous year and she ended up as a municipally-owned cattle ship on the Thames.

The screw steamer **Raven** was sold in September 1881 and was replaced on the Liverpool service by the **Bear**. The **Bear** had been displaced from the Londonderry route by the **Mastiff**. Thus, by the end of 1881 there were six new screw steamers running two services per night from Belfast and Glasgow, the **Owl** and the **Bear** were on the Liverpool overnight service, the **Mastiff** was running from the Clyde to Londonderry and one of the smaller steamers on the Belfast run fitted in the two crossings to Larne each week.

A small cargo steamer, the **Limpet**, with limited saloon and steerage passenger accommodation, was commissioned in 1882, designed for use on any of the services where cargo demand was high. She was also deployed on an irregular cargo service between Belfast and Ardrossan, but was also often used for the twice weekly Larne service. The hazards of winter navigation by such small ships in the North Channel were illustrated a few years later by a report in *Lloyd's List*, 23 February 1888:

> Ardrossan. The master of the **Limpet**, with a general cargo and passengers, reports having four plates on starboard side, between bridge and poop, crushed in by sea seven miles off Corsewall on Saturday evening.

The Burns' directors looked again at Ardrossan in 1882, realising that a passenger service between that port and Belfast had considerable potential with its shorter mileage. But there was the problem of the Ardrossan Shipping Company, with Robert Henderson acting as agent for the company at Glasgow. Originally created in 1844 as the Ardrossan Steam Navigation Company, the service was designed to encourage through business from Glasgow via the Ardrossan and Johnstone Railway (later the Glasgow and South Western Railway) and the proprietors of Ardrossan Harbour (see Christian Duckworth and Graham Langmuir, *Clyde and Other Coastal Steamers*, for company history). The Ardrossan company commissioned two new ships, the **North Eastern** and **North Western** in the winter of 1877/1878. Although smaller than the new Burns' ships on the Belfast route, they could manage 13 knots and offered the same creature comforts to saloon passengers that Burns did.

The main attraction to Burns was that these two ships, coupled with the train connection to and from Glasgow, offered a faster journey time than its own overnight train and mail boat via Greenock. Inevitably, a favourable offer was made by G & J Burns to the proprietors of the Ardrossan Shipping Company and, in 1882, Burns acquired a new route and two ships to run it, the **North Eastern** being renamed **Seal** and the **North Western** becoming **Grampus**. These ships provided a daily service, except Sunday, leaving Ardrossan at 11.05 pm and arriving Belfast at 05.00 am, and leaving Belfast at 6.30 pm. Return fares were interchangeable with the steamer from Greenock to Belfast. Burns at last had what it had always cherished – a route on which a daytime return steamer might yet succeed. The **Buzzard** was the outcome. She was ordered from J & G Thomson and was distinctive in that she was the first Burns steamer to be built with a steel hull, although the technology was tried, tested and accepted long ago by other ship owners, and the first to have two propellers. She was equipped with two powerful compound engines intended to provide a speed approaching 16 knots. The *Glasgow Herald*, Monday 17 March 1884, reported on her launch:

> The steel twin screw steamship **Buzzard** was launched from the building yard of Messrs Jas. & Geo. Thomson at Clydebank on Saturday forenoon. The vessel was named by Lady Edith Adeane, and among the company present were: Captain Adeane, RN, of HMS **Shannon**; Captain Pryce of the Board of Trade; Mr Clelland Burns and Misses Burns, Mr J C Burns, Major and Mrs Savage; Mr and Mrs George P Thomson… This beautiful vessel has been built for Mr John Burns to conduct a special daylight service between Scotland and the North of Ireland, and her salient points will be great speed with light draft of water, whilst everything is being done to conduce to exceptional comfort of passengers, so as to make the passage between the two countries essentially easy and agreeable.

The **Buzzard** was advertised to sail on her maiden voyage from Ardrossan at 9 am on 14 July 1884, with a scheduled arrival at Belfast at 1.45 pm. Just 45 minutes later she was intended to leave Belfast for Ardrossan on a very tight turn round, but with a more relaxed return journey of six hours and a scheduled arrival at Ardrossan at 8.30 pm. The **Buzzard** never quite achieved the fast morning schedule, although she tended to make up time on the return trip, after a late departure. Nor did the service ever attract many day trip passengers. Potential customers were not too enamoured by the idea of nearly twelve hours travelling for the sake of 45 minutes ashore at Belfast; the service was withdrawn at the end of August. Nevertheless, the same schedule was advertised to start again on 1 July 1885 with a 9.05 am departure. Again, the service was terminated at the end of August. The service was not resumed in 1886 and the **Buzzard** was sold the following year. She could not easily have been deployed on the existing overnight passenger services as she had only a few staterooms to offer and was essentially a day boat. The **Buzzard** did find some employment on extra cargo runs but spent a great deal of time laid up and out of steam. One of the critical issues of the ship was that she had a short length to breadth ratio of 6.5, a feature not conducive to speed as the vessel would drag water along behind her. The idea of a fast efficient and well patronised daylight return service was becoming an expensive pipe dream, but it would not be long before more cash was poured into it (Chapter 13).

An important new addition for the fleet was launched in December 1887. This was the steel-hulled passenger cargo steamer *Hare*. She was the first Burns ship to be equipped with a triple expansion steam engine, more efficient machinery than the ubiquitous compound engine. She also had a steam driven dynamo providing electricity for lighting in the accommodation spaces, the 'tween decks, and at strategic points on deck. The *Hare* was designed to support any of the Clyde services to the north of Ireland; she was used mainly to serve the Belfast route, but was also seen at Londonderry as a relief for the *Mastiff* during overhaul periods, and between Ardrossan and Belfast.

Meanwhile the *Owl* and the *Bear* continued to serve Liverpool and the Belfast service was still enjoying its six-ship line up, shared with the twice weekly Larne service. Londonderry was largely in the hands of the *Mastiff*, while the *Seal* and *Grampus* maintained the Ardrossan route and the *Limpet* served where and when she was needed. The fleet was modern and well equipped; the safety record of the company was second to none, and Burns ships were highly regarded by shippers and the travelling public alike.

Typical silver plate engraving on G & J Burns tableware in the early twentieth century.

CHAPTER 13

DAYLIGHT AT LAST

Experience with the ***Buzzard*** on the fast daylight crossing had provided a number of lessons. The first was that a long and thin profile was essential for speed, a ratio approaching, if not exceeding, 8.0 being desirable (***Buzzard*** had a mere 6.5). The practicalities of placing a heavy engine amidships into a screw steamer with such a narrow beam would lead to an unnecessarily deep draft to lessen the metacentric height, with consequent inefficiency and excessive rolling. The solution was to revert to paddle propulsion, in which a relatively shallow draft could be attained with an appropriate low profile engine design. The designers and draughtsmen set to work on such a concept, and a firm order for a new fast paddle steamer was placed by Burns in autumn 1888 with Fairfield Shipbuilding & Engineering Company at Govan.

Glasgow Herald Saturday 2 March 1889 reported under the headline 'New daylight service between the Clyde and Belfast: the Royal Mail steamer ***Cobra***':

The launch today of the Royal Mail steamer ***Cobra***, from the yard of Fairfield Shipbuilding and Engineering Company, signals an important development of the service between the Clyde and Belfast, as it is intended by Messrs Burns, the well-known ship-owners, for whom she has been built, to establish a new daylight service between the Clyde and Belfast. The vessel, which is of a particularly handsome model, was designed for the service by the late Sir William Pearce, Bart., and is to attain a speed of 19 knots or about 22 statute miles per hour. The speed will enable the vessel to complete the round voyage in one day. Passengers leaving Glasgow by train at 8 o'clock in the morning and joining the steamer at Ardrossan at nine will arrive in Belfast about one o'clock. In the capital of the North of Ireland they will be able to spend three hours in the pursuit of business or pleasure and may leave the port on the return voyage at 4 o'clock, arriving in the city at nine o'clock. This is in marked contrast to the earlier passages from the Clyde, and indeed those made at the present time…

The principal dimensions of the ***Cobra*** are as follows: length overall 275 ft; between perpendiculars 265 ft; breadth of beam 33 ft.; depth to promenade deck 22 ft. 6 ins.; and the gross tonnage will be about 1,000 tons. She has been built under Special Survey of Lloyd's Registry to Class A1 for Channel service, and also under the Board of Trade for the passenger certificate. The vessel has three decks. The promenade deck extends from stem to stern. On the main deck the principal apartments of the ship are situated. Forward the sides of the ship are carried up to the promenade deck, which provides an extensive shelter for steerage passengers. This space is lighted by circular side lights. This protected promenading space is continued in two wide corridors right aft, passing alongside the engine and funnel casings. Abaft the paddle boxes the skies are left open above the line of the main deck bulwarks for the purpose of giving light and ventilation to the main deck saloon which forms a special feature of the vessel. Outside of the saloon there is a sheltered promenading space for first class passengers, whilst along the sides are folding seats.

The saloon is large and beautifully furnished, and is lighted by square windows on both sides with skylights above. It is finished in solid mahogany relieved with gold decorations, and the upholstery is of fine crimson velvet. A departure is made in the arrangement which partakes considerably of the characteristics of the American style. There are along the sides partitions dividing the saloon into a large number of alcoves, each fitted with reading or writing tables. These are sure to be much appreciated. At the forward end of the saloon there is a private stateroom, elegantly furnished, for distinguished travellers. Convenient to the saloon is a cloak room, etc., and close by a pantry, with an elevator to the cuisine on the lower deck. Occupying the whole after part of the main deck is a large smoking room, furnished in oak and upholstered in brown morocco, and fitted with seats, revolving chairs and tables. Adjoining is a refreshment bar. The apartments are such as to make the smoking room a favourite retreat. The dining saloon practically takes up the part of the lower deck aft of the machinery and boilers. Access is provided by a wide stairway from the main deck, and light is admitted through two wells, besides the large number of circular side lights. The dining tables are arranged along either side, accommodation being provided for about 100. This room is finished in oak with maple panels. Forward of the dining saloon, on the same level, is a ladies' boudoir, fitted in walnut with satinwood panels, whilst the upholstery is in old gold. It is fitted with folding berths. There is lavatory accommodation in connection with it. On the lower deck forward there are two large rooms fitted up – the one for male and the other for female steerage passengers – and both apartments are arranged for the travellers dining. The mail and parcel post rooms are also located in this part of the ship.

The seamen etc., are accommodated in the extreme forward end on both the main and lower decks, whilst the officers' rooms are on the sponsons. The captain's room adjoins the wheelhouse on the promenade deck, and above them, stretching between the two paddle boxes, is the bridge, from which the ship is controlled. All the fittings here are of 'arguzoid', which gives the whole a bright appearance. Although the vessel is to be engaged in a daylight service, an installation of the electric light has been provided, a good-sized dynamo being placed in the engine room. The matter of ventilation has also been attended to, so that the comfort of the passengers has been carefully considered in all respects. The same remark applies to the even more important consideration of safety. The vessel is divided into nine separate water-tight compartments by bulkheads extending to the main deck, while four large life boats are carried.

The vessel will be fitted with a set of compound diagonal engines, having two cylinders and surface condenser. The indicated horse power will be about 4,000. The diameter of the high pressure cylinder is 50 ins., and of the low pressure cylinder 92 ins., the piston stroke being 5 ft. 6 ins. The high pressure cylinder is placed above the low pressure cylinder, and both are fitted with slide valves which are worked by the usual double eccentrics and link motion, and reversed by one of Messrs Brown Brothers' steam and hydraulic reversing engines. Nearly the whole of the castings, with the exception of the cylinders and valves and the condenser, are made of cast steel, so as to combine the greatest possible strength with lightness. The forgings are also made of Siemens Martin steel. The paddle shafts, cranks and crank pins are made of Vickers steel, and the shafts and pins are made hollow so as to reduce the weight. The water for condensing the steam will be circulated through the condenser by a centrifugal pump driven by a separate engine. The paddle wheels have feathering floats, which, together with the paddle arms, feathering rods etc., are made of steel. Steam is supplied from four single-ended boilers, 14 ft. 6 ins in diameter, by 10 ft. 3 ins. long, each having three of Fox's corrugated furnaces. The boilers are adapted for 110 lb. working pressure, and will be placed in two separate compartments forward and abaft the machinery. They have been constructed of steel, under the Special Survey of Lloyd's and the Board of Trade. The auxiliary machinery is to include Muir & Caldwell's steam steering gear and capstans forward and aft. In addition to the gear there will be two wheels aft for use in emergencies.

The vessel will present a very smart appearance. She will have two funnels and two light pole masts carrying fore and aft sails, giving them a rakish appearance.

Builder's model of the paddle steamer *Cobra* (1891) at the Riverside Museum, Glasgow.

In the event it was decided to use Gourock as the Scottish terminal rather than Ardrossan, where there were issues with connecting trains and potential tidal restrictions. Besides, Gourock would give access to specialist engineering skills should they be needed. On trials the *Cobra* attained the target 19 knots. G & J Burns took over management of the ship, although ownership was retained by the builders, ready to take her maiden voyage from Gourock on 15 June 1889; 'Grand pleasure sailing Glasgow to Belfast and back in a day for 6/- by the new and magnificent paddle steamship *Cobra*'. She left on time at 08.50 am and arrived in Belfast at 1.45 pm. Two and a quarter hours later she headed off down Belfast Lough, arriving at Gourock shortly after 8 pm. The *Cobra* quickly attained a following of businessmen who left Glasgow by train at 8 am and were home again in Glasgow by 9.40 pm. There were few tourists wanting to make the trip other than on peak summer Saturdays and during the Glasgow Fair Holidays, but tourist traffic did increase during the season.

During that first year it was realised that a number of aspects of the design of the *Cobra* conspired against her. The first was her cumbersome machinery, which required constant maintenance to ensure that it could be run at full speed and was very demanding in the volumes of steam needed to do this. The second, a follow on from the first, was that she was very demanding on coal.

As the summer progressed, a series of meetings were held between Burns' Superintending Engineer and the engineering design team at Fairfield Shipbuilding & Engineering Company. Both sides agreed that they could do better, and Fairfield especially was keen to show that it could improve on the design as a showcase for its own engineering prowess. The outcome was that the shipbuilder agreed to retain ownership of the *Cobra* and that a new and improved paddle steamer was commissioned by Burns to run the daylight service in time for the 1890 season. The design specification included a shallower draft and more efficient but powerful machinery. Some commentators have unfairly described the *Cobra* as a failure. However, all parties agreed at the time that the *Cobra* was an important ray of daylight towards the new service and that she was a huge success in her role in developing the daylight service; it is in this light that she should be remembered.

The **Cobra** was renamed **St Tudno** and placed on a Liverpool to North Wales service in the following year. At the end of the season she was sold to German owners.

The new steel-hulled paddle steamer, constructed in the winter of 1889/1890, was the **Adder**. She was longer than her predecessor by 15 feet, allowing a shallower draft to be attained. Her machinery was a much more simple, but even more powerful, compound diagonal engine which had cylinders of 51½ inches and 93 inches diameter and a stroke of 72 inches. Her design speed was 19½ knots. She afforded all the creature comforts of the **Cobra** and more. She was launched on 7 April with her machinery and much of the fitting out completed. On trials she managed just over 20 knots and was duly handed over to her owners. Negotiations for connecting train services to Ardrossan had come to nothing and the new ship was placed on the daylight service based at Greenock (**Cobra** had been based at Gourock).

The **Adder** took her maiden voyage on 14 June 1890. She was scheduled to leave Greenock at 8.45 am and was due alongside at Belfast at 1.30 pm. She left Belfast at 3 pm and was due back at Greenock before 8 pm. The return fare, including connecting trains from and to Glasgow, was set at 12/6d saloon and 6/- steerage. On Fridays and Saturdays a popular weekend excursion was offered for 15/-, returning on Monday afternoons. The **Adder** quickly became popular with the business fraternity, as well as tourists wanting to visit Belfast for the weekend. The engines of the **Adder** were much more robust than the complex machinery that had been installed in the **Cobra**, and they were able to see the season through without any significant modification or repair. The service was suspended for the winter at the end of September. The daylight era had finally arrived and, it would seem, was here to stay.

The Ardrossan to Belfast day passenger steamer **Adder** (1890) at speed in the Clyde; note the after funnel is caked in soot.

[W Robertson & Co]

Considerable kudos was attracted to the Fairfield Shipbuilding and Engineering Company on the back of the **Adder**'s performance. The theme was developed with two speculative builds, the **Koh-i-Nor**, which was commissioned in 1892, and the **Royal Sovereign** the following year, both being mortgaged into the Thames excursion trades. They were not cheap to build, the **Koh-i-Nor**, for example, came at the price of £50,900, almost double the cost of other new excursion steamers working in the Lower Thames. Nevertheless, all this effort paid off handsomely and orders were received for three 21 knot paddle steamers from the Dutch cross-Channel operators Zeeland Steamship Company.

The Burns' services were sustained throughout 1890 much as before. The **Owl** and the **Mastiff** maintained the Liverpool route in conjunction with M Langlands & Sons, while the **Bear** also featured from time to time. There were still two departures from both Belfast and Glasgow each night, except Sundays, for the overnight mail service. The service to Larne was twice weekly, with departures from Glasgow scheduled for Mondays and Thursdays at 4 pm, returning Tuesdays and Fridays at 9 pm. The Londonderry boat left Glasgow at 4 pm on Wednesdays and Saturdays and returned at 6 pm on Mondays and Thursdays.

Sir George Burns died on 2 June 1890, having been made a Baronet in the previous year at the age of 94. His business interests had spread far beyond shipping and he was the largest shareholder in the Glasgow & South Western Railway, which company also operated steamers on the Clyde. His eldest son, John Burns (Lord Inverclyde), inherited his title and had long since taken over the managerial responsibilities of his father including those of ship owners G & J Burns as a senior director, in addition to his managerial responsibilities for the Cunard Steamship Company. John's brother, James Cleland Burns (Cleland being his mother Jane's maiden name), was also widely involved in Clyde shipping circles and was one time chairman of the Glasgow Shipowners' Association. Sir George Burns had lived out his retirement at Castle Wemyss, the family home at Wemyss Bay. Following his father's death, John began to pass on the management of the various companies to his son, George. George Arbuthnot Burns (Arbuthnot being his mother Emily's maiden name) followed as Cunard chairman, as well as being a partner and director of G & J Burns.

The **Adder** resumed her daylight schedule in 13 June 1891 under the banner 'Grand Pleasure Sailings'. During her winter hibernation she had been refurbished and her machinery had been extensively overhauled, ready for the new season. An expensive ship such as this in use for only four months of the year suggests that, no matter what profit she reaped in the summer, the overall annual economic margins on the service were inevitably low. Comparison with the margins on the Clyde services to Belfast and Londonderry, the longstanding service to Liverpool, and the **Grampus** and **Seal** running year round from Ardrossan to Belfast, must have compelled the directors to question the merits of the daylight service on more than one occasion.

The second passenger sailing on the overnight Glasgow and Greenock to Belfast route was dropped at the end of the summer season in 1890. Thereafter a single overnight mail run was made from both Ardrossan and Glasgow via Greenock and return, nightly except Sundays. This allowed the smaller steamers, in a supporting role, to concentrate on cargo loading and discharge without attention to the strict departure schedules adhered to by the mail ships; in effect two ships were still employed on the route each way and each night.

Grouse (1891) on trials in the Firth of Clyde.

[Caird & Company Limited]

The Glasgow to Larne service received new tonnage when the **Grouse** was delivered by Caird & Company in summer 1891. She was equipped with a modern triple expansion steam engine of modest horsepower. Functional, with a single hold forward and one aft, she had no passenger spaces and was devoted to cargo carrying, along with stalls for 80 head of cattle and a dozen horses. She had the now common side doors, allowing livestock to walk straight aboard ship onto the main deck. There was a small deckhouse over the machinery space aft of the funnel and the open bridge was built up on framework overlooking the forward hatch and livestock pens. Functional she may have been, efficient she certainly was, and the little cargo steamer **Grouse** stayed in service with the company for the next 33 years. She was only broken up in 1948, her final role being under the name **Lochdunvegan**, working for David MacBrayne Limited.

In 1892 the **Adder** resumed her daylight service to Belfast on 1 June. She emerged newly painted with polished brasswork and everything clean and shiny ready for the new season. Her funnels, instead of the drab black of her owners had been repainted in an attractive buff-yellow colour with a black top. This year, instead of being based at Greenock she worked from Ardrossan, agreement over connecting train services from and to Glasgow having finally been made. Day return fares from Glasgow were 12/6d saloon and 6/- steerage. The new timetable offered a departure from Ardrossan at 10am arriving at Belfast at 1.35 pm. The **Adder** then left Belfast at the later time of 4 pm, allowing passengers to arrive at Glasgow at 9.30 pm on the special train up from Ardrossan. The morning departure from Glasgow Central Station left for Ardrossan at 9.05 am.

The mail ship departure from Ardrossan was scheduled to depart at 11.10 pm arriving at Belfast at 5.00 am, with a departure from Belfast at 9.30 pm. The Greenock sailing was timed at 9.55 pm, arriving at Belfast at 4.30 am, with departure from Belfast scheduled for 9 pm. The Liverpool service was offered at 15/- for a ten day return, 3/- steerage, with departures on Tuesdays, Thursdays and Saturdays from both Greenock and Liverpool, timed for a variety of departure times in the afternoon and evening. Most services at Liverpool received passengers at Clarence Dock, but some embarked passengers at the new Princes Landing Stage.

The **Hound** (1893) on the Clyde.
[Mike Walker collection]

The next new ship for the Belfast service was the much acclaimed steamer **Hound**, launched on 1 March 1893 from the yard of Fairfield Shipbuilding and Engineering Company at Govan. Her main role was on the Belfast Mail Service, which kept her busy throughout much of the year, although she was initially based at Ardrossan, working alongside the **Seal**. She was also used on the Clyde to Londonderry route, particularly in the winter months. The arrival of the **Hound** on the Ardrossan station in early summer 1893 allowed the **Grampus** to transfer to the Glasgow to Londonderry service.

During the autumn overhaul periods in 1893, both the **Grampus** and the **Seal** had their topmasts removed to provide sufficient air-draft for negotiating bridges on the new Manchester Ship Canal. From the opening of the Canal, the pair inaugurated a new passenger and cargo service between Glasgow and Manchester, with a maiden departure from Glasgow on 2 January 1894 by the **Grampus**. She arrived at Manchester Pomona Docks at 9 pm on the following day, after a prolonged nine hour passage up the Canal. This service was considered to be outwith the existing agreement between MacIver, Langlands and Burns for a shared and coordinated service from the Clyde to Liverpool and was soon a cause of great contention. That being so, Langlands soon followed suit and also ran up to Manchester.

With both the **Grampus** and the **Seal** absent from Ardrossan, one of the larger mail steamers was deployed at Ardrossan to work alongside the **Hound**, although there was much interchanging of vessels between the two routes thereafter.

Considerable change also came to the Liverpool route at the end of December 1895. MacIver (Liverpool and Clyde Steam Navigation Company) was particularly irritated by the new Manchester services operated by Burns and Langlands outside the Clyde/Mersey Conference, and elected to withdraw from the agreement. Robins and Tucker summarised the issues at stake:

> The problem stemmed from the various new cargo services operated by the Liverpool to Glasgow Conference members which were now running into Manchester. G & J Burns and the Glasgow & Liverpool Royal Steam Packet Company had collaborated closely since the first price war was resolved between them in the early 1840s. In later years, C MacIver & Company, Henry Lamont & Company and R Gilchrist & Company had all joined the club with agreed sailing schedules and rates across the five companies. But once the cargo services commenced between Glasgow and Manchester the rules no longer applied and commercial pressure meant that each company could take what it may. The consequence was an inevitable division between the five companies so that each reduced its rates to undercut the others. Bulk trades were the preserve of Gilchrist and Lamont in the original agreement but now Langlands and Burns vied with each other to pick up whatever trade was on offer, particularly bulk salt shipments from Runcorn destined for the chemical industry in Glasgow. In hindsight, each company must have quickly realised the need to negotiate with the others, but the damage was done and the negotiating table seemed far away.

The steamers **Owl** and **Bear** were transferred to MacIver as that company's share in the joint business, and the new steamer **Spaniel** and the **Mastiff** were then deployed by Burns on the Liverpool service. The Clyde Shipping Company's steamer **Cumbrae** was chartered in for a few trips until the **Spaniel** was ready. In May 1896 the **Mastiff** was replaced by the **Pointer**, an identical sister to the **Spaniel**. Fares were cut to 10/- cabin and 5/- steerage and all concerned lost a great deal of revenue.

The **Pointer** and **Spaniel** were a considerable advance on their predecessors on the Liverpool service. As before, a deck house on the poop contained the dining saloon and there was a smoke room in a deckhouse amidships. Two and four berth staterooms were located on the main deck aft beneath the promenade deck. Inside cabins had the

The *Pointer* (1896) alongside Donegall Quay at Belfast with the railway steamers beyond.

[Mike Walker collection]

The *Pointer* (1896) as she appeared in later life as the *Lairdsvale* for Burns & Laird Lines Limited.Limited.

novel feature of skylights penetrating the promenade deck, while outside staterooms enjoyed conventional portholes. As was now customary, steerage class was accommodated forward.

On 12 May 1896 a fire broke out in bales of cotton aboard **Spaniel** as she was being unloaded at Glasgow. The bales were quickly discharged onto the quay and the fire dealt with. In the meantime, considerable water damage occurred to the remaining cargo in the hold, but a serious threat to the ship had been averted.

The **Lamprey** was replaced by a new cargo steamer, **Ape**, in 1898. The **Ape** was a near sister of the **Grouse**, completed seven years earlier. The pair was of almost identical dimensions, although the new ship was two feet deeper than the **Grouse**. The **Ape** took up duties between Glasgow and Belfast in support of the mail service.

The cargo vessel **Ape** (1898) had a straightforward utilitarian appearance.

Two new ships, **Magpie** and **Vulture**, were built for the Ardrossan mail service and commissioned in summer and autumn 1898. They were single screw vessels and their triple expansion steam engines sustained a service speed not quite up to the 16 knots that they had attained on trials. The saloon passenger arrangement, inaugurated with the **Pointer** and **Spaniel**, included the dining saloon in a deckhouse on the poop and smoking room in a deckhouse amidships. Unlike that pair, the **Magpie** and **Vulture** had the sleeping berths arranged amidships rather than beneath the poop, an arrangement that was followed in all subsequent passenger ships built for the company. The arrival in service of the new ships allowed the **Hound** to return to the Clyde and Belfast Mail Service.

The **Lairdsrock** (1898), formerly the **Vulture**, boarding passengers at Greenock in the early 1930s.

Charles Waine wrote of the **Magpie** and **Vulture**:

> They were built by A & J Inglis for Burns and spent much of their time running between Ardrossan and Belfast initially. Dimensions were 265 x 33.6 x 15.9 feet and speed about 14 knots. Fidded topmasts were a feature of several Burns vessels. They appear flush decked but had poop, bridge and forecastle with hatches in deep wells with cargo doors. All the crew were forward below the forecastle which was reserved for steerage passenger seats. Only the captain was below the bridge. About 120 passengers were accommodated in 2, 3, 4 and 5 berth cabins on both the main and lower decks with the dining saloon on the poop deck. The main deck between the masts was given over to cattle stalls arranged around the hatches and either side of the boiler casing with 14 to 20 horse stalls. The after hatch was wide so a cattle gangway could be used to reach the lower deck which was also used for cattle stalls. About 254 cattle could be carried, exact number depending on the number of horse stalls required. The 'H' like structures above the side doors allowed a cargo gangway to be raised or lowered as the ship was loaded or unloaded when a tidal berth was being used. The horizontal bar was held by pins through the holes in the uprights. Cargo was wheeled or carried aboard and the derrick used to lower the goods into the hold or hoist it out if discharging.

Lloyd's List Monday 23 July 1900, reported an unfortunate collision of sisters:

> Belfast 22 July. Steamer **Dromedary**, of and for Glasgow, from Belfast, was run into yesterday evening outside Belfast Lough by steamer **Alligator**, of and from Glasgow, for Belfast. Both steamers have come to Belfast with considerable damage, chiefly above the water-line. **Dromedary** carried full complement of Glasgow Fair passengers returning home, of whom two were killed and about 50 injured. **Alligator** apparently had no passengers…

The **Dromedary** had her hand rail carried away from the stem to the bridge but was able to sail for Glasgow that evening. The **Alligator** had her steering gear badly damaged and a hole cut into a plate above water, and remained at Belfast until repairs could be carried out. Presumably the respective masters were called to account by the Burns' Directors in due course, but, given the fog that prevailed that morning, both will no doubt have been exonerated.

Lairdsgrove (1898) was built as *Magpie* for the Glasgow to Belfast service.

[DP World]

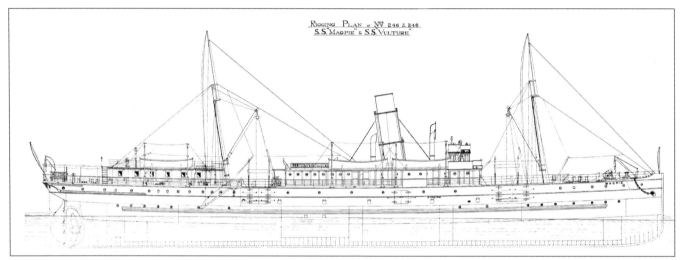

Builder's rigging plan for the *Magpie* (1898) and *Vulture* (1898).

[A & J Inglis]

The *Magpie* (1898) departing the Broomielaw with a full load of passengers.

Excerpts from *The Engineer*, Volume 64, 2 July 1897, pages 11-14:
The *Pointer* and *Spaniel*

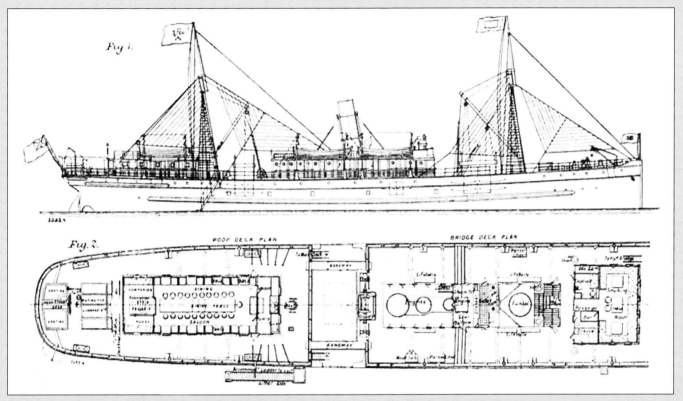

The screw steamers *Pointer* and *Spaniel*: Figures 1 and 2 from *The Engineer*, Vol 64, p11

We illustrate this week two coasting steamers, the **Pointer** and **Spaniel**, built by Messrs. A and J Inglis, Glasgow, for Messrs G and J Burns' trade from the Clyde to Belfast, Liverpool, and Manchester. The vessels are of special interest, as they embody a most successful effort at solving the difficulties of combining satisfactory sleeping and dining accommodation without both being unpleasantly associated. As is shown in Fig. 2, the saloon is on the poop deck, while the state-rooms are on the deck below, and thus there is no chance of sleeping passengers being disturbed by convivial suppers, or of unhappy voyagers being irritated by air laden with sweet suggestions of culinary successes. Another point which contributes to the comfort is the steadiness of the ships, alike as regards vibration and sea motion; and it should be stated that deep bilge keels have been fitted to reduce any tendency to roll. The principal dimensions of the ships are as follows:

Length of keel and forecastle 250 ft.
Breadth of beam (moulded on hurricane deck) 32 ft.
Breadth of beam, main deck 32 ft.
Breadth of beam, 'twoon decks 33 ft.
Depth, moulded 17 ft. 2 ins.
Sheer, forward 4 ft. 6 ins.
Sheer, aft 1 ft. 9 ins.
Length of poop from rudder post 70 ft.
Length of steerage from stem 27 ft. 6 ins.
Length of forecastle deck 47 ft.
Length of hurricane deck 77 ft.

The first-class passengers are accommodated in the poop in a large number of state-rooms, arranged as in the Atlantic service, while a large and handsome cabin with all conveniences is set apart for ladies. In the cabins are patent folding iron berths with spring and chain mattresses. A large lavatory with toilets and urinals is fitted up for the use of first-class passengers on this deck. The dining saloon is placed in a house on the poop deck, and is 42 ft. 6 in. by 15 ft. and 6 ft. 9 in. high. It is fitted up in a handsome manner, with dining tables, chairs, and sideboards. One feature of the design is the arrangement of side tables and settees, enabling parties of two or three to dine comfortably together. The saloon is well lighted by a skylight overhead, and by large rectangular brass-framed windows along each side.

A commodious pantry is placed at the fore-end of the saloon, with doors giving direct communication to the dining saloon. Abaft the saloon is the companion-house, covering the stair leading to the state-rooms. On the bridge deck (Fig. 2), a well-lighted and well-ventilated smoking-room is fitted up with tables and sofas all round. It is 17 ft. 3 in. by 21 ft. 3 in. Adjoining this room is the bar and larder. The captain's cabin is placed in this house also, and is well furnished. Indeed, both Captain MacVicar and Captain McLaren speak highly of the equipment, in every sense, of the *Pointer* and *Spaniel*.

Under the forecastle deck an open steerage is fitted up for the shelter of deck passengers, and a portion of the main deck is set apart for their use, while a bar and refreshment room is fitted up under the forecastle. The crew are berthed in a forecastle below the steerage, and the officers on the main deck, partly forward and partly amidships. The engine and boiler casings are carried up above the bridge deck to the boat deck overhead, thus providing an uninterrupted and protected promenade as shown by Fig 1. There are six boats, two being cutters. The galley is placed in the boiler casing; while the same section shows that the engine-room is lighted by a large teak skylight on the top of the casing, and also by round lights with brass frames, having outside polished flanges in the sides of the casings.

The main engines are of the triple-expansion type, with the starting platform on the port side. The cylinders are 24 in., 39 in., and 64 in. in diameter respectively by 42 in. stroke. The piston clearance at the top is ¼ in. and at the bottom ¾ in. The cylinder barrels are 1¾ in. thick at the sides and 2¼ in. at the bottom, with 1¾ in. liners. The covers are 10 in. deep, and the casing covers 6 in. deep. The pistons, 10 in. deep, are of cast iron and hollow, with strong ribs, the high pressure being provided with Ramsbottom rings, while the intermediate and low pressure have Lockwood and Carlyle's fittings. The rods are 6¼ in. in diameter with 7½ in. collars, the guide block faces being of Babbitt metal, with 250 square inches of surface. Indeed this metal has been used throughout. The connecting rod is forged, 8ft. long, 7 in. in diameter at the bottom, and 6 in. at the top, with 3¾ in. and 2¾ in. pins.

There are two single-ended boilers, 14 ft. 6 in. mean diameter and 12 ft. 4 in. mean length working at a pressure of 150 lb. Each has three of Morison's corrugated furnaces, 8 ft. long by 3ft. 5 in. mean diameter, rounded up at the back tube plates to 4 in. radius, and with separate fireboxes to each furnace. The furnace bars are in two lengths 3 ft. long by 4 in. deep by $7/8$ in. The grate surface is 117 square feet. The tubes are of iron, 3¾ in. in diameter, the spaces between the nests of tubes being 12 in. The total heating surface is 3754.8 square feet. The funnel is 7 ft. 6 in. in diameter, and is 39 ft. high above the hurricane deck. There is an unusually tall donkey boiler in the stokehold between the two bunkers. It is 20ft. high, and 6ft. 6 in. mean diameter. The firebox and water tubes are of Staffordshire BB iron and welded. The water tubes are 10 in. in diameter and 7/16 in. thick. There are 10 of them. The water space around the firebox is 10 in. at the top and 6 in. at the bottom, the total heating surface being 271.3 square feet, and the grate area is 22.5 square feet. The working pressure is 70 lb., suitable for the deck machinery. The boilers are worked with two of Weir's pumps which are located in the engine room, while Galloway's ash hoist is in use. There is capacity for 100 tons of coal; and here it may also be said that 40 tons of water ballast can be carried in the after hold and fore peak, while galvanised iron water tanks are provided for carrying 3000 gallons of fresh water.

The ship is lighted throughout by an electric plant consisting of Chandler's engine, driving direct at 300 revolutions a Harvey dynamo, compound wound fitted on a platform in the engine-room. The wiring is on the lead and return system, the conductors being protected by galvanised iron tubes ¼ in. thick, where they pass through the bulkheads or boiler rooms or coal bunkers; otherwise they are in wooden casings. There are about a hundred 16 candle-power lights and seven 50 candle-power lamps for holds, mooring, and lighting passengers' gangways.

CHAPTER 14

G & J BURNS LIMITED

The daylight dream had at last been achieved, with the fast paddle steamer **Adder** maintaining a day return service from Ardrossan to Belfast in the summer months. But the board of directors did have one last great aspiration – to run a direct service between the Clyde and Dublin. For many decades Burns' advertisements had signalled that its nightly mail service was to 'Belfast, Dublin, Londonderry…' without specifically alluding to the train journey involved in getting to places beyond Belfast. Indeed Burns would have broken into the Dublin trade long ago, had it not been for the existing direct service maintained by Burns' friends in the Laird Line. The other company involved in the service was the Dublin and Glasgow Steam Packet Company, and it might be, just might be, that one day the acquisition of this company would enable Burns to trade to Dublin. The Burns' perception all along, was that the two existing direct services between the Clyde and Dublin were eroding business from its own service to Belfast with its fast rail connection on to Dublin city centre. However, by the 1900s this assertion was probably no longer the case, as there was sufficient business available for all three parties.

John Burns died in 1901, as reported in *The Engineer*, 15 February 1901:

> The announcement of the death of Lord Inverclyde, or Sir John Burns of the Cunard Line, as he was still best known by in shipping and commercial circles came unexpectedly to many, though it was known locally that for some time his health had been failing. At a late hour on the evening of the 12th inst. he died in his seventy-second year at Castle Wemyss, on the Firth of Clyde.
>
> As head of the Cunard Steamship Company for many years, he was a leading personality, well known and esteemed throughout the shipping world, and when, in 1897, on the occasion of the Diamond Jubilee, he was raised to the peerage as the first Lord Inverclyde, it was generally recognised that a graceful and deserved tribute had been paid, not only personally, to one of sterling worth and eminence in the mercantile world, but also impersonally to the mercantile marine, and the Cunard Line, for the part it had played in bridging the Atlantic and linking together the English speaking peoples of the New and the Old World. Lord Inverclyde was the elder son of Mr George afterwards Sir George Burns, Bart, of Wemyss Bay – who with his brother James, in 1824, founded the shipping concern of G & J Burns…

With the newer and larger ships on service, there were fewer relief cargo and cattle sailings needed to and from Belfast. In 1901 there were three departures each week for Manchester and Liverpool, but just one to Liverpool in the winter season. Langlands and MacIver had pooled their resources on the Clyde to Liverpool service with a joint service operated by **Princess Louise** and **Owl**. A period of consolidation set in, with all of the independent operators on the Irish Sea suffering from the increasing competition being presented by the railways. Rail services between Liverpool and Glasgow, and more particularly between London and Glasgow were increasingly set at barely economic fares. At sea, the railways also offered through train connections for their own cross channel-steamer services from both Stranraer and Heysham. Although Burns, and for that matter the Laird Line, still held on to the majority share of the Irish traffic between Scotland and Ireland, they did so by cutting passenger fares and freight rates to an extent that the long term viability of some of the services was questionable. Nevertheless the Belfast and Londonderry routes continued to thrive throughout the 1900s and the Glasgow to Manchester service remained profitable, although passenger numbers were beginning to decline by the middle of the decade.

Minor incidents continued at a regular frequency much as before. For example, the **Mastiff** lost her rudder in 7 fathoms of water on 30 January 1902 when she was just four miles from Londonderry, and a tug was dispatched to tow her to port. Two days later, she was towed back to Glasgow by the Clyde Shipping Company tugs **Flying Swallow** and **Flying Elf**. In the following year, on 14 January 1903, fire broke out in one of her cabins while on passage to Londonderry, then three months later, on 13 April, the **Mastiff** had to turn back to Glasgow with a damaged propeller. The **Seal** and **Grampus** never quite got to grips with the confines of the Manchester Ship Canal and had various scrapes and collisions, not least when the **Seal** collided with the stonework of Trafford Bridge, above the main docks, on 3 September 1902, sustaining considerable damage to her bows.

The worst incident was the collision between the **Ape** and the dredger **Greenock** on the evening of 18 November 1902. The **Ape** was outward bound for Belfast and the collision occurred below Gourock, approaching the Cloch Lighthouse. The dredger was struck on the port quarter, filled rapidly and sank; one young crew member of the dredger was killed. The **Ape** put back with her bow severely damaged.

On 28 December 1904 a private limited liability company was registered under the name G & J Burns Limited. With effect from 1 January 1905, this new company incorporated all the assets, staff and goodwill of the partnership previously trading under the name G & J Burns and its strap line 'The Royal Mail Line'. Ships were registered under the new company during January.

John Burns' son George Arbuthnot Burns died in 1905 at the age of just 44. Although he had married, he had no children. His obituary in *The Engineer* 13 October 1905, included:

> A large number will hear with regret of the death of [the second] Lord Inverclyde, which occurred at Wemyss Castle, Wemyss Bay, last Sunday afternoon... His Lordship belonged to a family which for more than four generations has held an honoured place in the city of Glasgow. It is in connection, though, with the Cunard Steamship Company that the name of Inverclyde is perhaps most intimately associated. His grandfather was one of the founders of the company, and it is due to the indefatigable energies of his father, the first Baron Inverclyde, that the Cunard Company holds its present influential position in the shipping world... Besides his connection with the Cunard Company, he was a director of the Glasgow and South Western Railway Company, and the Clydesdale Bank and he was also a partner and director of the firm of Messrs G & J Burns, Limited.

The *Hound* was given a major overhaul in 1905 when her passenger accommodation was refurbished and a number of single berth cabins were installed. Thereafter, she returned to duty on the Ardrossan to Belfast route.

During 1905 the Marine Superintendent had been charged by the directors to look into the possibility of a turbine steamer replacing the paddle steamer *Adder* on the seasonal daylight service between Ardrossan and Belfast. George Burns, as Chairman of Cunard, had been a great advocate of turbine propulsion on the Atlantic, the outcome being the turbine steamer *Carmania*, commissioned in 1905. The Cunard enthusiasm for the new technology no doubt coloured the decision by Burns to pursue the concept of a direct-drive turbine steamer. This type of ship had triple screws, with the central screw driven by the high pressure turbine and the wing propellers by two lower pressure turbines. Direct-drive turbine steamers had been delivered for use on the English Channel from 1903 and were a great success; an almost standard speed of 21 knots, three knots more than that of the *Adder*, was readily achievable.

The technology had been championed by William Denny, in collaboration with Charles Parsons, the inventor of the marine turbine engine, with the Clyde steamer *King Edward*. She was launched on 16 May 1901 and delivered to the Turbine Steamer Syndicate as a showpiece for maritime interests to take note of. There followed a spate of orders from the railway companies and then, in 1905, two independent companies also placed orders – G & J Burns Limited and the General Steam Navigation Company. The outcome was the famous Burns steamer *Viper* and the long-distance Thames excursion steamer *Kingfisher*.

The *Viper* was a most worthy successor to the *Adder* when she took her maiden voyage on 1 June 1906. She had a service speed of 22 knots and had a passenger certificate for 1,700, twice that of the *Adder*, and required a crew of 60 including catering staff. On her bow was a coiled golden viper as a reminder to all that she was the third generation snake to run the Belfast daylight service, the first, of course, being the chartered paddle steamer *Cobra*. The *Viper* was equipped with an exceptionally powerful reverse turbine which, with a large and effective bow rudder, allowed her to turn while off Carrickfergus, on her approach to Belfast, and reverse up Belfast Lough all the way to her berth at Donegall Quay. Her accommodation was described in the *Glasgow Herald* 12 March 1906, following her launch two days earlier, when she was christened by the third Lord Inverclyde, James Cleland Burns and Chairman of G & J Burns Limited:

> The Fairfield Shipbuilding and Engineering Co. (Limited), Govan, launched on Saturday the steamer *Viper*, the first turbine steamer built for the Royal Mail services of J & G Burns (Limited) between Scotland and Ireland. Built to the highest class of Lloyd's Special Survey and to the Board of Trade requirements, she has been carefully planned and luxuriously furnished to ensure that passengers may enjoy the highest degree of comfort on board. The guaranteed speed at sea is 21 knots so that the passage between Ardrossan and Belfast will only occupy about 3¾

The *Woodcock* (1906) was renamed *Lairdswood* in 1929 and the following year sold to the Aberdeen Steam Navigation Company Limited and renamed *Lochnagar*, as seen here.

Turbine steamer *Viper* (1906) leaving Ardrossan for Belfast; from a postcard with the caption 'On board the Royal Mail turbine *Viper*'.

hours. Such expeditious travelling between these points has hitherto been impossible. The *Viper* has four decks, lower, main, spar and promenade. In accordance with the most recent developments in passenger steamer design, the accommodation for first class passengers has been placed forward of the centre of the vessel. On the spar deck are the entrance halls, the smoke, tea and staterooms, information office and luggage spaces. Forward on the main deck are the first class main saloon and first class ladies' cabin, while aft on the same deck are the second class saloon and second class ladies' cabin. The dining saloons are on the lower deck, that for first class passengers is 78 feet long and has seats at tables for 125 diners, while the second class saloon has seats for 105.

The many advantages of placing the first class accommodation forward will be obvious, while the only disadvantage – that of exposure to wind and weather upon stormy days – has been very simply obviated by the fitting of a shelter bulkhead of steel, built from side to side of the vessel, between the spar and promenade decks, at the fore end of the first

*Interiors of the **Viper** and **Woodcock** from a contemporary brochure.*

class accommodation. The shelter promenade behind this bulkhead, under the boat deck and in the lee of the range of deckhouses on the spar deck, will afford a very comfortable protected space for the enjoyment of fresh air and scenery even on the stormiest days. Six lifeboats will be placed on the promenade deck...

The *Adder* was sold for further service in Argentina as the *Rio de la Plata*. The *Viper* was a huge success. She was scheduled to leave Ardrossan at 10 am and arrive at Belfast at 1.20 pm. Her return sailing from Belfast was at 4 pm, with a scheduled arrival at Ardrossan timed for 7.15 pm. The *Viper* closed the season at the end of September, leaving the *Partridge* in charge of the overnight mail service from Ardrossan.

The *Partridge* had been commissioned for the Ardrossan night service, along with sister *Woodcock*, in July 1906. The pair were designed for the Glasgow to Belfast service, but commissioned on the Ardrossan route where they replaced the *Gorilla* and *Hound*. The *Woodcock* was relieved at Ardrossan in due course to serve on the Glasgow to Belfast route. The pair had first class accommodation for 140 passengers, while the steerage accommodation was more generously appointed than in any previous steamers in the Burns' fleet. The ships had 315 permanent stalls for cattle and four more for horses. They were built by John Brown at Clydebank at a cost approaching £100,000; the *Woodcock* was launched on 10 April 1906 and the *Partridge* on 23 May. The triple expansion machinery drove a single screw to provide a service speed of 15½ knots.

Lurcher (1906) was specifically built for the Manchester to Glasgow route and as such had short masts to negotiate the fixed bridges on the Manchester Ship Canal.

The **Vulture** was back in service in August and, on 17 August, her master, Captain Paterson, reported that when entering Ardrossan Harbour and making for Montgomerie Pier at 3 am, the vessel had refused to answer her helm and had struck the pier stem on, severely damaging her bows, and no doubt shaking up her sleepy passengers.

Two new sister ships, with the complementary canine names **Lurcher** and **Setter**, were also completed in 1906. These were smaller ships than the mail ships, with a large deadweight but without any saloon passenger accommodation; there was, however, space available for steerage passengers. They were distinctive looking, as they had stump masts only marginally higher than the funnel in order to provide sufficient airdraft for light passages in the Manchester Ship Canal, and of course there were no saloon passenger deckhouses. They came into service in September and October respectively, allowing the **Seal** and **Grampus** to stand down. The latter pair was sold to owners in Trieste, who gave them the names **Eleni** and **Elda**. The **Lurcher** and **Setter** together maintained the Glasgow to Manchester service with an unrivalled regularity until nearly the end of the Great War.

In November 1907, severe gales disrupted many of the Irish Sea services. On 13 November, the much delayed departures of **Partridge** and **Vulture** sailed from Belfast bound for Ardrossan. Both lost a number of cattle in the night due to the heavy seas, while numerous other animals were injured in a protracted and fierce storm. Although the **Vulture** was able to enter Ardrossan harbour during a lull in the wind, the **Partridge** had to continue on to Greenock without attempting the narrow harbour entrance at Ardrossan.

The **Partridge** inaugurated a set of summer cruise excursions in July 1907 with a departure from Belfast for Campbeltown, allowing passengers ashore to visit Machrihanish on the train. These excursions were repeated in subsequent years, and the duties were shared with the **Woodcock**.

During autumn 1907 the **Alligator**, then on the Belfast service, was sold in anticipation of a new ship then being constructed at the Pointhouse yard of A & J Inglis Limited. This was the **Redbreast**, of similar design, although slightly narrower in the beam, to the earlier **Woodcock** and **Partridge**, which had been built by John Brown at Clydebank. The **Redbreast** was a little faster than the **Woodcock** and **Partridge**, with a service speed of 16 knots. The new ship was launched on 18 April 1908 and was ready to take up service on the Belfast station in time for the peak summer season.

Despite the expense of all this new building, there was money in the bank and Lord Inverclyde looked again at the prospects of acquiring a direct Dublin service from Glasgow. He looked at the Dublin & Glasgow Steam Packet Company, he scrutinised that company's finances and its business prospects, faced as it was by competition from the better equipped Laird Line, and he weighed up a commercial value for the Dublin company, its assets, staff and goodwill. With a generous bonus to that figure he went to the Board of the Dublin company and proposed his purchase offer, which was debated and finally agreed in Dublin at the end of January 1908.

The Dublin company had a long and proud history, as described by Charles Waine:

The **Redbreast** (1908) lying alongside a busy Broomielaw with a Clyde paddle steamer setting off 'doon the watter'.

> Dublin & Glasgow SP Co, Dublin, was formed in 1836 by Dublin interests, to operate between Dublin and Greenock using both steamers and sailing vessels, hence it was originally known as the Dublin & Glasgow Sailing and Steam Packet Co. Early wooden paddle steamers ran in opposition to the St George SP Co but in later years worked in association with the City of Cork SP Co, thus avoiding serious competition, so Dublin and Glasgow schedules were extended to Cork and Bristol. An iron paddle steamer, **Vanguard** 700 gross tons, was delivered in 1843, but twenty years later, **Lord Gough** 705 [tons gross, built] 1863 and **Lord Clyde** 705 [tons gross, built] 1863, iron paddlers, financed by selling older steamers to the Confederates, set new standards of speed on the Dublin-Greenock route, now reduced by an hour to 12 and a quarter hours, later ships cutting the time still further. Vessels were henceforth to be named after titled folk; and eventually in 1892 the firm was christened Duke Line. The last paddle steamers being **Duke of Leinster** 736 [tons gross, built] 1870 and **Duke of Argyll** 809 [tons gross, built] 1873, with **Duchess of Marlborough** 402 [tons gross, built] 1874 introducing the screw. She was a cargo ship with limited passenger accommodation, but some more impressive passenger ships followed, the firm operating a nightly Dublin to Glasgow service, via Greenock, in competition with the Laird Line, but the latter only handled a third of the trade. Vessels like **Duke of Rothesay** 1,226 [tons gross, built] 1899 and **Duke of Montrose** 1,389 [tons gross, built] 1906 proved popular...

Thus, in 1908, Burns acquired four screw passenger and cargo steamships with the takeover. Burns formed a new subsidiary company, the Burns Steamship Company Limited, to own and manage the ships. The ships were the

Duke of Gordon, **Duke of Fife**, **Duke of Rothesay** and **Duke of Montrose**. The eldest of the four ships had been commissioned in 1885 as the **General Gordon** and only given the more corporate name **Duke of Gordon** ten years later. On entering the Burns Steamship Company Limited, her original triple expansion steam engine was replaced by a new set built and installed by D Rollo & Sons at Liverpool. At the same time she adopted the name **Wren**. She was rarely deployed on the long distance Glasgow to Dublin service, but was a useful vessel for some years in the Londonderry and Liverpool trades.

Duke of Fife (1892) was renamed *Sparrow* when she was under the ownership of the Burns Steamship Company.

Duke of Rothesay (1899) was renamed *Puma* in 1908 and adopted the new corporate name *Lairdsford* in 1929.

Tiger (1906) at Gairloch on one of many seasonal West Highland Cruises she carried out between 1922 and 1926. She was built originally for the Duke Line as the *Duke of Montrose*.

Tiger (1906) ended her days as the cattle carrier *Lady Louth*.

The next oldest member of the Burns Steamship Company was the **Sparrow**, formerly the **Duke of Fife**. She had been completed in 1892 and was distinguishable from her peers at a distance by her low profile and exceptionally tall funnel. The **Puma**, ex-**Duke of Rothesay**, was a more modern vessel built much along the lines of the Burns steamers of that era, with first class accommodation aft and steerage in the forecastle. The fourth ship to be acquired by Burns, the **Duke of Montrose**, was renamed **Tiger**; she was an almost new vessel that had been launched by the Caledon Shipbuilding and Engineering Company at Dundee on 10 April 1906, just two years previously. Initially the three newer ships were used to maintain the Clyde to Dublin service, supported by the **Magpie**. Sailings were at 2 pm from Glasgow and 7.30 pm from Greenock on Mondays, Wednesdays, Fridays and Saturdays, with departures from Dublin at 6.30 pm. The service remained under the independent banner, The Duke Line, until 1 May 1908 when it was fully integrated within the G & J Burns brand.

In 1909, the **Dromedary** was withdrawn and sold, followed by the **Gorilla** in 1913. The third ship in this trio, which had been completed for the Belfast mail service in 1881, the **Alligator**, had already been disposed of in 1907. The ships deployed by the company were all now thoroughly up to date and modern, as noted in an address by Lord Inverclyde: 'Things were advancing very quickly in ship-owning and they found that they had to be continually replacing their steamers with newer boats'.

In 1910 the **Sparrow** was sold to Greek owners, and ships were drafted in to fill the gap from other services, as best they could. A new steamer was ordered for the Dublin route but she was not commissioned until 1912. A new cargo and cattle steamer was commissioned in 1911, with the name **Sable**. She was principally employed on Clyde to Belfast services, sometimes calling at Larne, but was available for other routes as demand required. She replaced the **Ape**, which was sold shortly after the **Sable** had been commissioned.

The new steamer for the Dublin route was the magnificent passenger and cargo ship **Ermine**, the pinnacle of ship design in the entire fleet of G & J Burns Limited. She was launched from the Govan yard of Fairfield Shipbuilding & Engineering on 15 June 1912. The **Ermine** was unique in the fleet in that she had twin engines driving twin screws, providing a service speed of 16 knots. The engines were rather special, being hybrid four cylinder triple expansion engines, each having a conventional high and intermediate cylinder, which then discharged into two low pressure cylinders. There were three boilers supplying steam at 175 pounds per square inch, one single ended and two double ended units. Another novel feature for a Burns ship was an enclosed wheel house, complete with centrifugal panes for clear vision.

Ermine (1912) served G & J Burns only briefly before war was declared on Germany.

The *Glasgow Herald*, 17 June 1912, reported on the launch of the vessel:

> A new twin screw passenger steamer – the **Ermine** – was launched on Saturday by the Fairfield Shipbuilding and Engineering Company, Govan, for Messrs. G & J Burns, Glasgow. The **Ermine**, which is intended for her owner's service between Scotland and Ireland, is 311 ft. in length overall, 40 ft. in breadth, and 17 ft. 6 in. in depth to the main deck. The vessel will have accommodation for 170 first class passengers in large two and four berth rooms on the promenade and hurricane decks, while there will be also a block of portable cabins on the main deck fitted, when required, for 30 additional passengers. The first class dining saloon will seat 70 people, and the seats are so arranged that they can be converted into sleeping accommodation for 25 passengers when required. At the after end of the hurricane deck there will be deck house cabins providing sleeping berths for 30 passengers. The second class accommodation will be aft on the main deck, and the whole of the midship section of this deck will be used as a second class promenade and shelter. The vessel will also have accommodation for a considerable number of third class passengers. The forward hold and the lower deck forward and aft are to be used for cattle or cargo, while the main deck amidships can also be used for this purpose when it is not required for passengers.
>
> The vessel was named by Mrs Spens, wife of Major General Spens, General Officer Commanding the Lowland Divisions of the Territorial Force. Among others on the launching platform were General Spens, Lord Inverclyde and Mr J A Roxborough, Directors of Messrs. G & J Burns…

The **Ermine** was commissioned in summer 1912, and was an immediate success with the travelling public on the Dublin route. During that summer the deployment of the fleet was broadly:

Viper – seasonal daylight return service between Ardossan and Belfast
Partridge, *Woodcock*, *Redbreast*, *Vulture*, *Hound* – variously employed on overnight sailings Glasgow (Broomielaw) and Greenock to Belfast and Ardrossan to Belfast
Magpie, *Puma*, *Tiger* – overnight service from Glasgow (Lancefield Quay) and Greenock to Dublin
Wren – Glasgow (Broomielaw) and Greenock to Londonderry
Spaniel, *Pointer* – Glasgow (Broomielaw) and Greenock to Liverpool
Lurcher, *Setter* – Glasgow to Manchester, with optional calls at Greenock, Ardrossan and Ellesmere Port or Runcorn

Sable was mainly deployed on extra cargo sailings to Belfast, while **Magpie** was often also used on this service.

The Dublin service was subject to all the vagaries of severe weather in the Irish Sea. On 9 January 1913 the **Ermine**, outward bound from the Clyde to Dublin, received a morse lamp signal from the inbound steamer **Puma**, while off Kildonan, on Arran, to the effect that the bridge of the **Puma** had been carried away in a storm during the night. Within the week, the **Ermine** had also been damaged by weather, sustaining considerable damage on deck. On a happier note, the **Ermine** undertook a number of summer cruises in 1913 and again in 1914, mainly from Greenock to Iceland, and in each case with a reduced passenger complement of just 80.

The summers of 1913 and 1914 were the last great days of tourist traffic between Scotland and Ireland; the political problems in Ireland were an increasing deterrent to visitors. Significantly, the political issues developing in Europe were of more concern, as the inevitability of war with Germany loomed large. On 4 August 1914, Britain declared war against Germany following that country's violation of Belgium's neutrality. The background to the war was complex and involved numerous failed interrelationships and declarations. Few people in Britain understood why their country was at war, although all championed the cause in the safe belief that the war would be over by Christmas. There was initially a sense of carry on regardless, although the **Viper** had already made her last sailing of the season between Ardrossan and Belfast on Friday 7 August and had retired to lay up at Glasgow on the following day; darkness once again fell over the daylight service. All other services by Burns remained as before, while Glasgow excursion sailings to lower Clyde ports continued unabated.

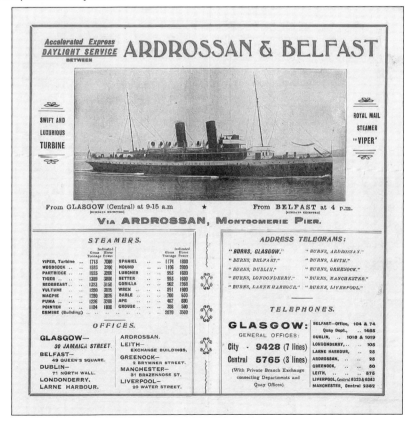

August 1912 Ardrossan to Belfast advertisement.

CHAPTER 15

WAR, PEACE, TAKEOVER AND MERGER

It did not take long for the priorities of war to disrupt the schedules offered by G & J Burns Limited. Burns proudly advertised in mid-November 1914 that 'Sailings between Glasgow, Belfast, Dublin, Londonderry, Manchester and Liverpool are being fully maintained except no passengers will be carried from Glasgow to Manchester on 12th and 17th'. By mid-December the Thursday sailings from Glasgow via Greenock to Dublin had been cancelled and the Tuesday sailing was direct to Dublin, without calling at Greenock. Despite all, the **Spaniel** and **Pointer** maintained the Ardrossan to Belfast service throughout the war, and the **Lurcher** and **Setter** did their best to maintain two sailings a week between Manchester and Glasgow. The former Duke Line ships **Puma** and **Tiger** were retained essentially on Dublin sailings but the pair worked on other routes from time to time as required. Most of the remaining passenger and cargo ships served where they were required, with a great deal of interchange between services.

Matters came to a head in late 1914, when the **Woodcock** was requisitioned as an armed boarding steamer HMS **Woodnut**. The following year, the **Partridge** was also requisitioned by the Admiralty and given the name HMS **Partridge II**, the **Redbreast** was taken into Government service, later becoming a fleet messenger, and the **Ermine** was designated fleet messenger No 22.

The **Woodcock** was requisitioned on 15 November 1914 for conversion to an armed boarding steamer. In 1915 she was renamed HMS **Woodnut** to avoid confusion with another ship of the same name. Converted for use as a fleet messenger she was sent to the Mediterranean to support the Gallipoli campaign alongside HMS **Partridge II**, **Redbreast** and **Ermine**. For a long while the **Partridge** ran a supply line between Malta and Marseilles. Both **Partridge** and **Woodcock** survived the hostilities and were returned to their owners in 1919.

The **Redbreast** was commissioned on 17 July 1915 as a fleet messenger. She first went to Mudros on 29 September and remained in support of the Gallipoli campaign throughout 1916 and into 1917. For part of the time she doubled as a decoy Q Ship, with a single gun concealed forward. On 15 July 1917 she was torpedoed and sunk by UC-38 in the Aegean Sea, while on passage from Skyros to the Doro Channel. Forty two members of her crew were killed.

The **Ermine** retained her name as a fleet messenger and, by 6 September 1915, had arrived at Mudros. She then commenced a shuttle service between Suvla Bay and Mudros carrying troops and military stores, loading alongside other ships and frequently discharging under fire. She continued this work throughout 1916 and into 1917. The end came on 2 August 1917, when she struck a mine in the Aegean Sea, laid by UC-23, while on passage from Stavros to Mudros carrying passengers and stores. She sank in position 40.39 N 23.34 E, with the loss of 24 lives.

The **Viper** was put to use as a transport in the English Channel. She was deployed mainly on trooping duties from Dover, and from 1916 was based at Southampton running to Le Havre as support to the British Expeditionary Force. The **Vulture** was also used as a transport, notably working between Aberdeen and Bergen towards the latter part of the war.

Donegall Quay, Belfast, in the 1920s with **Sable** (1911) or **Coney** (1918) extreme left and **Woodcock** (1906) beyond. Further away are the railway ferry to Heysham and the Liverpool boat.

[Mike Walker collection]

The *Moorfowl* (1919), as built, before extensive reconfiguration in 1926, wearing Burns & Laird Lines colours, after she was transferred into the fleet in 1920 from the City of Cork Steam Packet Company.

With so many of its ships away from home from 1915 onwards, G & J Burns was put in charge of three ships belonging to the North of Scotland & Orkney & Shetland Steam Navigation Company in 1916: the **St Margaret**, **St Magnus** and **St Sunniva**. However, none was used on Burns' own Irish Sea services and the three saints took up duties at the behest of the newly appointed Shipping Controller. The **St Margaret** was torpedoed and sunk while so employed, when on passage from Lerwick to Iceland on 12 September 1917. The ship was 30 miles east of the Faroes when the torpedo exploded in the bunkers. The **St Margaret** sank in just four minutes, killing four men out of the crew of 22. The survivors were able to sail the lifeboat to Hillswick, a distance of 150 miles, which they accomplished in three days.

The frustrations of trying to operate the Irish services in wartime are illustrated by a company advertisement placed in the press on 1 May 1917: 'No sailings for passengers from Glasgow or Greenock to Dublin (North Wall) today, Tuesday 1st May; No sailing from Glasgow or Greenock to Londonderry tomorrow 2nd May'.

Safety was by no means assured on the Irish Sea services. On 13 September 1918 the **Setter** was on a routine voyage from Manchester to Glasgow in the approaches to the Clyde. At 5 am, at a point 6 miles north west by north of Corsewall Point, the **Setter** was torpedoed by submarine UB 64 without warning and sank rapidly. Fourteen of the crew were lost.

A new cargo steamer was delivered in 1918 with the name **Coney**. She had been authorised by the Shipping Controllerate as a replacement for vessels lost in the war or absent on official duty. The **Coney** was launched from the yard of A & J Inglis Limited at Pointhouse on 14 February and was in service shortly afterwards. She had a long forward hold, suitable for the carriage of steel components destined for the shipbuilders at Belfast and, like her near sister the **Sable**, completed before the war, had two hollow steel masts topped by the extractor vents from the cattle deck, and fidded topmasts. A new passenger and cargo steamer was also authorised and A & J Inglis launched this ship with the name **Moorfowl** on 1 April 1919. The ship did not come up to expectations and, following a dispute between builder and ship owner, the vessel was rejected as not fit for purpose; the ship was later completed for the City of Cork Steam Packet Company Limited, a recently acquired subsidiary of Coast Lines Limited, as the **Killarney**.

G & J Burns emerged from the war with a hugely depleted fleet and a long list of sea-going and shore-based staff that had lost their lives during the war. Some of these men and women had taken up arms with the forces, only to lose their lives for what they had deemed a just cause. As Britain stumbled to its feet in 1919, there was a realisation that it was a nation greatly depleted, with war debts owing to the United States that it could barely hope ever to repay, and with a fragile industrial base that was focused on military hardware.

Burns very boldly put the refurbished turbine steamer **Viper** on Belfast duties in 1919. However, given the poor state of the economy, and exacerbated by the increasing political problems facing Ireland, the ship lost money as the season progressed. The **Viper** was withdrawn before the end of September due to a rail strike. During the winter lay-up period the ship was put on the sale list as unlikely to be profitable in the foreseeable future. In July 1920 she sailed to Douglas as the Isle of Man Steam Packet Company's **Snaefell**. Years later in 1945, the **Snaefell**, former **Viper**, was sold for demolition at Port Glasgow, although she was not broken up until 1948. Her main saloon staircase was removed and incorporated into the rebuilding of HMS **Wellington**, during that vessel's conversion in 1947/48 into a Livery Hall as HQS **Wellington**. She is now the home of the Honourable Company of Master Mariners, and is moored alongside the Victoria Embankment in London – complete with the staircase from the **Viper**.

Earlier in the year, on 16 January 1920, the Burns' directors accepted a generous offer made by Alfred Read of Coast Lines Limited to purchase the entire holdings of G & J Burns Limited and the Burns Steamship Company Limited. Read had money to spend and Burns was faced with poor prospects in post-war Britain, prospects that were compounded by

One of four cargo steamers originally destined for M Langlands & Sons was the *Gorilla* (1921) seen here in later life as the *Cambrian Coast*.

[B&A Feilden]

The *Lurcher* (1922) was also originally destined for M Langlands & Sons but was launched in G & J Burns' colours as *Lurcher* but transferred to Coast Lines as *Scottish Coast* in 1923.

the poor state of affairs in Ireland that were fast developing into civil war. As with the Laird Line before it, the sale was something of a fire sale and the true value of the company was not realised. For the moment G & J Burns Limited was left largely to its own devices under the watchful eyes of its new parent.

One week after the takeover had been agreed a replacement for the *Setter*, lost in the war, was launched from the yard of William Beardmore & Company Limited at Dalmuir. However, she did not stay long in the fleet; Coast Lines management transferred her to the British & Irish Steam Packet Company later in the year with the new name *Lady Kildare*. The *Lurcher* was transferred with her and was given the name *Lady Meath.*

Three new ships were authorised by the Coast Lines board in 1920 for M Langlands & Sons Limited, a subsidiary of Coast Lines Limited since 1919. The first two ships were allocated the names **Princess Dagmar** and **Princess Caroline** but, at their launches from the Harland & Wolff yard at Govan in October and December 1920, they were christened **Redbreast** and **Gorilla** and placed under the management of G & J Burns. They were delivered in February 1921 and February 1922 respectively and placed on the Clyde to Liverpool service. They were single deck ships with one hold, the engines and accommodation being right aft. The third ship in the trio was launched from the Inglis yard at Pointhouse as G & J Burns' **Lurcher** in September 1922. Two similar ships were ordered from A & J Inglis, but the orders were quickly transferred from G & J Burns Limited to the British & Irish Steam Packet Company and allocated the names **Lady Limerick** and **Lady Olive**. The order for the **Lady Limerick** was cancelled but the second ship was eventually delivered in September 1922 under the allocated name.

In 1920, the daylight Ardrossan to Belfast service was put in the charge of the chartered steamer *Graphic*, owned by the Belfast Steamship Company. However, she was an overnight steamer and could not offer her day passengers the comforts that the *Viper* had done previously. The *Graphic* opened the season on 1 July and was advertised to sail from Ardrossan at 10 am, but took 1 hour 20 minutes longer on passage than the *Viper* and arrived at Belfast, Albert Quay, at 2.40 pm. The return sailing was still scheduled for 4 pm, the steamer moving over to Donegall Quay to pick up her passengers, and getting passengers back to Glasgow by connecting train at 9.45 pm. The direct mail service from the Broomielaw, at 9 pm, no longer called at Greenock, while the overnight service from Ardrossan to Belfast departed at midnight. The Dublin service had three sailings a week, on Mondays, Tuesdays and Thursdays from Glasgow at 12 noon, leaving Greenock at 4.35 pm. There were two sailings to Londonderry, one on Wednesdays, normally at 12 noon, and one on Saturdays at 6 pm. Liverpool still had its regular nightly service run, once again in collaboration with M Langlands & Sons.

The *Graphic* finished the daylight season on Saturday 14 August and was returned to her owner. The charter was carried out at a loss and Burns resolved not to run a daylight service next year. As it happened it was a good choice; there was a three month long coal strike from April 1921, which necessitated careful rostering of steamers and slow running to conserve fuel stocks.

A pair of new passenger and cargo ships was designed for the British and Irish Steam Packet Company Limited's overnight Liverpool to Dublin route, and built by Ardrossan Dockyard Limited. That company had previously designed and built the **Ardmore**, which it had delivered to the City of Cork Steam Packet Company in 1921. This same design was used as the foundation for the two new sister ships which were ordered by British & Irish early in 1922. However, in December the orders were reallocated to G & J Burns Limited and the first ship was thereafter known as **Ermine**. She was launched in March 1923 without a name, and then in May 1923, the anonymous ship, then at the fitting out stage, was christened **Lady Louth** and registered under the ownership of the British & Irish Steam Packet Company. Her consort followed suit and became the **Lady Limerick.**

The *Irish Times*, 7 July 1923, reported:

> The SS **Lady Louth**, passenger, cargo and cattle steamer, the latest addition to the fleet of the B&I, has just been completed by her builders, the Ardrossan Dry Dock & Shipbuilding Company. The vessel will be used in the Dublin/Liverpool trade and is 276 feet in length, has a gross tonnage of 1,881 tons and enjoys the latest improvements of her class, including wireless. The structural arrangements and scantlings are to the requirements of Lloyd's highest class and Board of Trade requirements for the safe carrying of passengers, and fitted with the latest life-saving appliances.
>
> Accommodation for 80 passengers is arranged on the bridge and promenade decks. The lounge, which is situated at the forward end of the promenade deck, is panelled and furnished in the Tudor period. The dining saloon, which seats forty persons, is tastefully panelled in light coloured oak. Spacious accommodation for 91 steerage passengers is provided in the poop space.
>
> Particular attention is given to the cattle-carrying arrangements, the ship having splendid capacity for the safe and efficient carrying of cattle and horses, which are to the latest requirements of the Board of Agriculture for the Irish cattle trade.

The new ship also offered some single berth cabins, hot and cold sea water baths, and a special sitting room for ladies, who also had a 'hairdressing establishment'. Burns had indeed missed out on re-establishing the overnight Glasgow and Belfast route. It would seem that the drop in passenger numbers to Belfast was such that the Coast Lines management believed the new sisters would fare better working from the Free State port of Dublin, at least until such time as the 'troubles' settled down. Even so, the **Lady Louth** was faced with a three month strike shortly after she started on the Liverpool to Dublin run, a direct result of the troubles.

On 25 July 1922 the names G & J Burns Limited and the Burns Steamship Company passed into history with the formation of the merged company Burns & Laird Lines Limited. The inevitability of the merger in terms of economic savings was considerable and the merged company was able to pool resources as well as draw on those of other members of the Coast Lines' empire. The G & J Burns ships that joined the merged fleet were: **Coney**, **Gorilla**, **Grouse**, **Hound**, **Magpie**, **Moorfowl**, **Partridge**, **Pointer**, **Redbreast**, **Sable**, **Vulture** and **Woodcock** and there were two from the Burns Steamship Company: **Puma** and **Tiger**. **Woodcock**, and **Partridge** remained the staple pair on the Clyde to Belfast route, **Puma** and **Tiger** ran mainly to Dublin, although **Tiger** was also used for seasonal cruises in the mid-1920s; the engines aft cargo ships, **Gorilla** and **Redbreast** served Manchester and Liverpool, along with the **Lurcher** which joined them after the merger.

In 1923 the advertised timetable offered the following:

Glasgow to Belfast, daily at 9 pm, 10.45 pm Saturdays
Ardrossan to Belfast, daily, except Saturday, train departs Glasgow Central 10.30 pm
Ayr to Belfast, cargo only, daily except Saturday
Ayr to Larne, cargo only, Mondays, Wednesdays and Fridays
Glasgow to Londonderry, daily, train to Greenock 9.40 pm
Glasgow to Sligo, Coleraine and Mulroy, cargo only, as required
Glasgow to Liverpool and Manchester, cargo only, twice weekly
Heysham to Londonderry, twice weekly

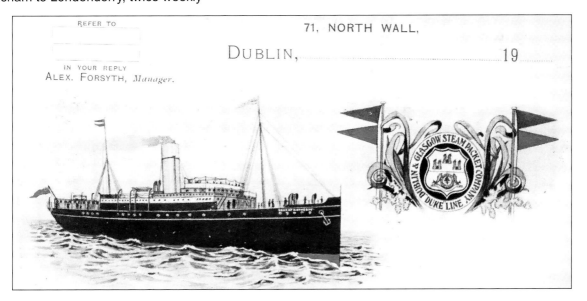

Duke Line postcard advising: **Duke of Rothesay** Sept 21st - 22nd 1906 Glasgow to Dublin; **Duke of Montrose** Dublin to Gourock, train to Glasgow. Sept 28th-29th 1906.

Ironically the **Lady Limerick** and **Lady Louth** did eventually come under Burns & Laird Lines Limited's management in 1930, when they served on the Glasgow to Belfast overnight route as the **Lairdscastle** and **Lairdsburn** respectively. Even the prototype of the group, the **Ardmore**, joined forces in 1930, operating the Glasgow to Dublin service as the **Lairdshill**. Many of the old Burns ships had very long careers: the Duke Line's **Duke of Rothesay** spanned 35 years in the Irish Sea trades and the **Vulture** of 1898 transferred to David MacBrayne as their 'cruise ship' **Lochgarry** in 1937.

Burns & Laird Lines struggled through the 1920s and early 1930s before a revival in trade allowed new ships. These were the **Royal Ulsterman** and **Royal Scotsman**, commissioned for the Glasgow to Belfast overnight services as replacements for the **Lairdscastle** and **Lairdsburn**.

Burns, Laird and their successor Burns & Laird are remembered for a variety of reasons by different people. A lasting memory is the smell of cattle, fried cooking and stale beer tainted with fuel oil, a scent that was guaranteed to make even the strongest stomach turn queasy once the ship was beyond the shelter of the Cumbraes.

A letter by Bridie Stevenson of Thornliebank, Glasgow, to *The Glasgow Herald* 9 February 2000, reminisced about the discomfort of travelling, other than first class, by Burns & Laird aboard the **Lairds Loch:**

> Every year at the start of the school holidays, my mother took us to Ireland on the Derry Boat. What memories I have! Mothers opening suitcases and placing their babies on top of the clothes to sleep, children lying sleeping on the floor and irate mothers trying to prevent the drunks lurching about from falling on top of them, toilet basins blocked with vomit which sloshed from side to side with the boat's movement, women on their knees saying the Rosary out loud in Gaelic and English as the boat heaved (usually around midnight passing Paddy's Milestone - Ailsa Craig) and oh the smell! Diesel and the strong smell of cattle permeated every corner of the ship (another reason for the vomit!).
>
> Passenger comfort would appear to have been an alien concept to Burns & Laird who crammed passengers on board just like the cattle. The seats were either canvas or hard wooden slats and the limited food available invariably added to the stomach upsets.
>
> My last trip on the Derry Boat was at the end of the Glasgow Fair in July 1962. I was now grown up and working, but still returned to Donegal every year for my holidays.
>
> As usual the boat was completely overcrowded. We could not even find standing room, far less a seat, and had to spend the entire 12-hour journey on the open deck - even in July the night air at sea is extremely cold! The next day I collapsed in the street and spent five weeks in hospital with pneumonia and pleurisy. I vowed never to go on that boat again!
>
> Since then my frequent trips to Donegal have been made via Cairnryan or Stranraer or by airline companies. However, if the 'Derry Boat' were to rise from the ashes, albeit a modern, comfortable version, then I know that I, along with thousands of other Scots with their roots in Donegal, would welcome the chance to travel direct from Glasgow to Derry again.

In the final throes of the Coast Lines group before it was taken over by P&O in 1971, a number of unit load ships were given rather uncharacteristic and seemingly non-corporate names. These included a **Pointer** and a **Spaniel**, even a **Buffalo** – where could such names have come from? We know don't we?

Revival of a G & J Burns name: the unit load motor ship **Pointer** (1956) heading down the River Ribble below Preston Dock on 7 April 1969.

[Nick Robins]

Lairdshill (1921) alongside the Anderston Quay at Glasgow on 27 July 1934. She was built for the City of Cork Steam Packet Company Limited as *Ardmore* and with sister *Kenmare* provided the prototype for the *Ermine* (1923) launched as *Lady Louth* and an unnamed hull launched as *Lady Limerick* (1924).

[F W Hawks]

Lairdscastle (1924) was commissioned as *Lady Limerick* for the British & Irish Steam Packet Company but was originally ordered for use by G & J Burns.

An advertising card for Burns & Laird Lines from the 1920s featuring the *Tiger* (1906).

Lairdsford (1899), one time ***Duke of Rothesay***, below Preston Dock in February 1934 on her way to the breaker's yard.

[D Cochrane]

The ***Vulture*** (1898) was renamed ***Lairdsrock*** in 1929 and was sold to David MacBrayne in 1937 and considerably rebuilt for summer cruises as the ***Lochgarry***.

[Linda Gowans collection]

The ***Hazel*** (1907) as ***Mona*** alongside.

REFERENCES

Campbell, Colin and Fenton, Roy 1999. *Burns and Laird.* Ships in Focus Publications, Longton.

Clague, Dick 2004. *Heysham Port: a Century of Manx and Irish Services.* Ferry Publications, Ramsey.

Duckworth, Christian and Langmuir, Graham 1967. *West Highland Steamers,* 3rd edition. T Stephenson & Sons Ltd, Prescot.

Duckworth, Christian and Langmuir, Graham 1977. *Clyde and Other Coastal Steamers*, 2nd edition. T Stephenson & Sons Ltd, Prescot.

Fairplay 1885 *Clydeside Cameos.* Reproduced in book form from the trade and shipping journal *Fairplay.*

Hyde, Francis 1975. *Cunard and the North Atlantic 1840-1973: A History of Shipping and Financial Management.* The MacMillan Press Limited, London.

Kennedy, John 1903. *The History of Steam Navigation.* Charles Birchall Limited, Liverpool.

Larn, Richard and Bridget (various dates for different volumes). *Shipwreck Index of the British Isles.* Lloyd's Register, London.

Liddle, Laurence 1998. *Passenger Ships of the Irish Sea.* Colourpoint Books, Newtownards.

MacHaffie, Fraser 1975. *The Short Sea Route.* T Stephenson & Sons Ltd, Prescot.

McNeill, David 1969. *Irish Passenger Steamship Services, Volume 1, North of Ireland.* David & Charles, Newton Abbot.

McNeill, David 1971. *Irish Passenger Steamship Services, Volume 2, South of Ireland.* David & Charles, Newton Abbot.

Newall, Peter 2012. *Cunard Line – a Fleet History.* Ships in Focus Publications, Longton.

Patton, Brian 2007. *Irish Sea Shipping.* Silver Link Publishing Ltd, Kettering.

Robins, Nick and Meek, Donald 2006. *The Kingdom of MacBrayne*, 1st edition. Birlinn Ltd, Edinburgh.

Robins, Nick and Tucker, Colin 2015. *Coast Lines Key Ancestors: M Langlands & Sons.* Bernard McCall, Portishead.

Robins, Nick and McRonald, Malcolm 2017. *Powell, Bacon and Hough, Formation of Coast Lines Limited.* Bernard McCall, Portishead.

Sinclair, Robert 1990. *Across the Irish Sea, Belfast-Liverpool Shipping Since 1819.* Conway Maritime Press Limited.

Trollope, Anthony 1878. *How the 'Mastiffs' Went to Iceland.* Virtue & Company, London.

Waine, Charles 1999. *Coastal and Short Sea Liners.* Waine Research, Albrighton.

Williamson, James 1904. *The Clyde Passenger Steamer 1812-1901.* James MacLehose & Sons, Glasgow.

APPENDIX 1

FLEET LISTS

All ships registered at Glasgow while in Laird or Burns and associated company ownership, unless stated otherwise. Gross tonnage is the original tonnage as built.

1) Laird and associate companies

Lewis MacLellan & Others, Londonderry Steamboat Company, Clydesdale Steamboat Company, Foyle Steamboat Company, St Columb Steamboat Company (1814-1835)

Name	LM&O Service	Gross tons	Comments
Britannia	1815-1829	73 net	Paddle steamer built by John Hunter, Port Glasgow, 1815, for Lewis MacLellan, Archibald McTaggart & others; 1821 lengthened by John Wood & James Barclay, Port Glasgow, ownership transferred to Britannia & Waterloo Steamboat Co; 1826 sold to Britannia Steamboat Co; wrecked in gale 13 October 1829 at Donaghadee.
Waterloo	1815-1825	90 net	Paddle steamer built by John Wood & James Barclay, Port Glasgow, 1815, as **Waterloo** for Lewis MacLellan, Archibald McTaggart & others; 1825 sold to James Morrison & others and renamed **Maid of Islay**; 1826 re-built by Wm. Simons & Co, Greenock, and sold to Claud Girdwood & Co, Glasgow; 1834 sold to John Stalker; 1835 sold under mortgage to William Guthie, Portobello, Andrew Greig, Newhaven, John Sword, Leith & Dugald Twiner, Leith; wrecked 27 October 1835 in Firth of Forth.
Argyle	1824-1826	72 net	Paddle steamer built by John Hunter, Port Glasgow, 1815, as **Argyle**; 1821 lengthened by 12 feet, registered by George Brown & Thomas Buchanan, Glasgow; 1826 sold to Argyle Steam Boat Co, Glasgow; 1828 sold to John Ritchie, **Glasgow**; 1828 sold to Andrew Drysdale & others, **Glasgow**; 1828 sold to Andrew Drysdale, John Mitchell & others, **Glasgow**; 1829 sold to John Mitchell & others, Alloa; 1843 scrapped.
Londonderry	1826-1835	102 net	Paddle steamer built by William Denny, Dumbarton, 1826, and launched as **Belfast**, completed for Londonderry Steamboat Co, Glasgow, as **Londonderry**; 1835 sold to William Quinn, Liverpool; 1841 scrapped.
Clydesdale	1826-1828	87 net	Paddle steamer built by McMillan & Hunter, Greenock, 1826, as **Clydesdale** for Clydesdale Steamboat Co, Glasgow; 15 May 1828 on fire during voyage from Glasgow to Belfast, run aground and later wrecked off Corsewall Light.
Foyle	1829-1844	225	Paddle steamer built by J Lang, Dumbarton, 1829, as **Foyle** for Foyle Steamboat Co, registered at Londonderry; 1835 passed to Glasgow & Londonderry Steam Packet Co; 1844 sold to Glasgow & Liverpool Steam Shipping Co; 1847 sold to William B Brownlow, William Hunt Pearson & George Holmes (Brownlow, Pearson & Co), Hull; 1853 sold to H T Cheeswright & W T Miskin, London; 1857 sold to C Joyce & another; 1859 sold to Zachariah C Pearson and converted to sailing ship, schooner rigged; 1861 sold to Thomas R Oswald, Sunderland; 1861 sold to Henry Briggs, Hull; 1861 sold to George C Beckett, Sunderland; 19 February 1862 wrecked at Sines, Portugal, in gale during voyage from Carthagena to Sines.
Saint Columb	1834-1851	238	Paddle steamer built by J Wood, Port Glasgow, 1834, as **Saint Columb** for St Columb Steamboat Co, registered at Londonderry; 1835 passed to Glasgow & Londonderry Steam Packet Co; 1851 sold to Robert P Stephens, Glasgow; 1860 Register closed, sold to a party in Liverpool and believed broken up.

Glasgow & Londonderry Steam Packet Company – Glasgow agent T Cameron & Company (1835-1867)

Name	G&LSP Co Service	Gross tons	Comments
Rover	1836-1852	350	Paddle steamer built by Wood & Mills, Dunglass, 1836, as *Rover* for Glasgow & Londonderry Steam Packet Co, registered at Londonderry; 8 December 1852 sank off Bengore Head, following a collision with the Fleetwood steamer *Princess Alice*, on voyage from Londonderry to Glasgow.
Londonderry	1841-1859	513	Paddle steamer built by Robert Steele & Co, Greenock, 1841, as *Londonderry* for Glasgow & Londonderry Steam Packet Co and registered at Londonderry; 1859 scrapped.
Rambler	1845-1846	620	Paddle steamer built by Robert Napier, Govan, 1845, as *Rambler* for Glasgow & Londonderry Steam Packet Co and registered at Londonderry; 16 June 1846 aground on Maidens Rocks, off the Antrim coast, on voyage from Liverpool to Sligo, and became a wreck.
Shamrock	1847-1879	714	Paddle steamer built by Caird & Co, Greenock, 1847, as *Shamrock* for Glasgow & Londonderry Steam Packet Co, registered at Londonderry; 1867 passed to Lewis MacLellan, registered at Glasgow; 1867 passed to Walter MacLellan, Lewis MacLellan, John Reid & Alexander A Laird; 1879 sold to Charles Gasquet, London; 1879 scrapped.
Thistle	1848-1858	653	Paddle steamer built by Robert Napier, Govan, 1848, as *Thistle* for Glasgow & Londonderry Steam Packet Co; grounded 9 December 1858 on Roughly Point, near Sligo, on voyage from Sligo to Glasgow, scrapped.
Laurel	1850	429	Paddle steamer built by Caird & Co, Greenock, 1850, as *Laurel* for Glasgow & Londonderry Steam Packet Co, but completed for Glasgow & Liverpool Steam Shipping Co; 1854 sold to Humphrey John Hare & Co as nominees for London & North Western Railway Co; 1867 scrapped.
Rose	1851-1867	489	Paddle steamer built by Robert Napier, Govan, 1851, as *Rose* for Glasgow & Londonderry Steam Packet Co; 16 July 1867 wrecked in Sligo Bay.
Myrtle	1854	405	Paddle steamer built by A Stephen & Sons, Kelvinhaugh, 1854, as *Myrtle* for Glasgow & Londonderry Steam Packet Co; 4 August 1854 wrecked on Baron Rock, near Sanda, off the Mull of Kintyre, on voyage from Glasgow to Londonderry.
Arbutus	1854-1866	292	Built by Thomas Toward, Newcastle, 1854, as *Arbutus* for London & North Western Railway Co; 1858 engines compounded, lengthened by 35 feet; 1865 sold to John Burns, James C Burns & Charles MacIver (G & J Burns); 1865 sold to Alexander A Laird; 1866 sold to Belfast Steamship Co; 1872 sold to James S Campbell, Dublin; 1872 new engines; 1880 sold to J M Inglis; 1883 sold to James Tulloch & Co, Aberdeen; 1885 sold to Robert Taylor, Dundee; 17 January 1890 wrecked on Goldstone Rock, Holy Island, Northumberland, on voyage from Dundee to Seaham.
Garland	1855-1867	283	Built by Robert Napier & Sons, Govan, 1855, as *Garland* for Glasgow & Londonderry Steam Packet Co; 1867 sold to William Middleton & William Polloxfen, Sligo, lengthened by 22 feet and renamed *Glasgow*; 1867 passed to Sligo Steam Navigation Co Ltd.; 1879 new engines; 1889 sold to Miss Lola Armstrong, Liverpool; 1890 sold to James M Allen, Halifax, Nova Scotia; 1890 sold to unknown Spanish owners and renamed *Emilio*; July 1891 wrecked at Yabucoa, Puerto Rico, on voyage from Ponce, Puerto Rico.

Bee	1857-1867	40	Lighter built at Greenock, 1857, as **Bee** for John Cameron & Lewis MacLellan; 1867 passed to Lewis MacLellan on the death of John Cameron; 1867 sold to Thomas McGowan & James Fernand; 1867 sold to William Middleton & William Polloxfen, Sligo; 1885 register closed, vessel at Sligo, a wreck, and unfit to proceed to sea again.
Myrtle	1859-1865	44	Paddle steamer built by Palmers Shipbuilding & Iron Co Ltd, Jarrow, 1859, as **Myrtle** for London & North Western Railway Co registered in name of William Whelan as nominee at Lancaster; 1865 passed to William Briggs, Morecambe, following the death of William Whelan, and sold to William E Hutchinson, & William P Price as nominees of Midland Railway Co; 1871 sold to John Jacobson, Joseph Wilson & John L Rodgett, all of Morecambe; 1873 sold to Robert Wilson, Morecambe; 1873 sold to Thomas Hunter & John Steen, Barrow; 1876 sold to Joseph Hunter; 1879 scrapped.
Thistle	1859-1862	387	Built by Laurence Hill & Co, Port Glasgow, 1859, as **Thistle** for John Cameron & Lewis MacLellan; 1862 sold to George Wigg, Liverpool, as blockade runner; 8 March 1862 aground off Battery Beauregard, while leaving Charleston, salvaged, sold to John Ferguson and renamed **Cherokee**; 8 May 1863 captured by USS **Canandiqua** and USS **Flag** and taken into US Navy as USS **Cherokee**; 1865 sold out of US Navy to Benjamin Vicuña Mackenna, Chile, for use as a warship in the war against Spain; 1872 converted to survey vessel; 1878: Sold to J F Sanchez, Chile; wrecked 25 August 1889, off Punta Guaban, Chile, on voyage from Melinka to Valparaiso.
Thistle	1863	472	Paddle steamer built by Laurence Hill & Co, Port Glasgow, 1863, as **Thistle** for John Cameron & Lewis MacLellan; 1863 sold to James A K Wilson, Liverpool as blockade runner; captured 4 June 1864 east of Charleston by USS **Fort Jackson** and taken into US Navy as USS **Dumbarton**; 1868 sold to Quebec and Gulf Ports Steamship Co, Quebec, as **Dumbarton**, new engines and renamed **City of Quebec**; 1 May 1870 sank at Green Island, Saint Lawrence, after collision with Allan Line's **Germany**.
Laurel	1863-1864	387	Built by A & J Inglis, Pointhouse, 1863, as **Laurel** for John Cameron & Lewis MacLellan; 1864 sold to Henry Lafone, Liverpool, as blockade runner and renamed **Confederate States**; 1865 sold to John Moody, Goole & Richard Moxon, Yorkshire, and renamed **Walter Stanhope**; 1865 sold to Goole Steam Shipping Co Ltd; 1868 lengthened by 27 feet, new engines; 1881 sold to William Cawthorn, London; 1882 sold to Peter Hutchinson, Glasgow, and renamed **Niobe**; 1883 sold to Peter Hutchinson & others; 1887 sold to A Mickilse, Bordeaux; France, and renamed **Bordelais**; 1888 sold to J & P Hutchinson, Glasgow and renamed **Niobe**; 28 November 1905 sank in Le Havre Roads, after collision with Booth Line's **Gregory**, on voyage from Glasgow to Rouen.
Myrtle	1865-1867 1871-1877	159	Built by Barclay, Curle & Co, Whiteinch, 1865, as **Myrtle** for Glasgow & Londonderry Steam Packet Co, registered in names of John Cameron & Lewis MacLellan as agents; 1867 registered in names of Walter MacLellan, Lewis MacLellan, John Reid & Alexander A Laird as agents; 1867 sold to Joseph C R Grove, London; 1871 sold to Alexander A Laird; 1872 passed to Glasgow & Londonderry Steam Packet Co and registered in names of Alexander A Laird, Walter MacLellan Lewis MacLellan & John Reid as agents; 1872 new engines; 1877 sold to Peter McGuffie, Liverpool; 26 January 1878 wrecked at Ballycastle Bay on voyage from Portrush to Liverpool.

Name	Service	Gross tons	Comments
Thistle	1865-1883	444	Built by A & J Inglis, Glasgow, 1865, as *Thistle* for Glasgow and Londonderry Steam Packet Co and registered in names of Trustees; 1877 new engines; 1883 sold to Whitehaven Steam Navigation Co; 1885 sold to Robert Tedcastle & Co, Dublin; 1897 sold to D & C MacIver, Liverpool; 1897 sold to Union Transport Co Ltd, Manchester; 1900 sold to Islander Steamship Co, Bristol; 1908 scrapped.
Irishman	1867-1878	270	Built by Blackwood & Gordon, Paisley, 1854, as *Irishman* for Alexander A Laird (Glasgow & Dublin Screw Steam Packet Co); 1858 managed by MacConnell & Laird; 1863 lengthened by 17 feet following sinking off Ardglass, Co Down; 1867 managed by Glasgow and Londonderry Steam Packet Co; 1873 passed to Alexander A Laird & Co; 1878 sold to Harland & Wolff; 1883 new engines, renamed *Dunmurry*; 1887 sold to Brodsky & Tweedy (Russian Sea, Land & River Co), Odessa, Russia, and renamed *Nadeshda*; no further details.

Glasgow & Dublin Steam Packet Co – McConnell and Laird (1851-1867)

Name	G&DSPC service	Gross tons	Comments
Northman	1851-1854	181	Built by Denny Brothers, Dumbarton, 1847, as *Northman* for Kirkwall Steam Navigation Co, Kirkwall; 1851 sold to Alexander A Laird (Glasgow & Dublin Steam Packet Co); 1854 sold to William Cameron, John Bell and John Patton, Glasgow; 1855 sold to Alexander M Dawson, Liverpool and used as transport ship at Balaclava; 1856 sold to John O Lever, Manchester; 1858 sold to Zachariah C Pearson, James Coleman and Edward Coleman, Great Grimsby, and resold to Henry T Watson, Hull, lengthened by 40 feet; 22 February 1859 sank off Flushing, Netherlands, on voyage from Antwerp to Goole, after collision with American ship *Union*.
Irishman	1854-1867	270	See *Irishman* above.

McConnell & Laird (1853-1867)

Name	M&L service	Gross tons	Comments
Prince of Wales	1853	500	Built by Tod & McGregor, Glasgow, 1842, as *Prince of Wales* for builders; 1843 sold to John Laidlay, Fleetwood, Thomas L Birley, Kirkham, Henry Smith, Preston, James Bourne, Liverpool, & others (North Lancashire Steam Navigation Co); 1843 sold to John Laidlay, Preston & Wyre Railway Harbour & Dock Co, Thomas L. Birley & others; 1849 sold to Preston & Wyre Railway Harbour & Dock Co, John Laidlay, Frederick Kemp, Fleetwood, Thomas L Birley & others; 1850 sold to Frederick Kemp, Preston & Wyre Railway Harbour & Dock Co, Thomas L Birley & others; 1853 sold to Frederick Kemp, Lancashire and Yorkshire & London and North Western Railway Co's (joint), Thomas L Birley & others; 1869 sold to Walter K Bayley, Birkenhead, and converted to sailing ship; 1869 sold abroad and renamed *Rose Brae*. Re-sold to Walter K Bayley, Liverpool; 1873 sold to William H Jones; 21 April 1875 sank at 44° N, 48° W, after striking an iceberg on 20 April, on voyage from Galveston to Liverpool.

Name	Service	Gross tons	Comments
Emerald	1855-1858	247	Built by R Napier & Sons, Govan, 1855, as *Emerald* for R Napier & Sons and sold to Robert Armour, Patrick J Mills, William Whyte, Alexander A Laird & James R Napier; 1857 sold to Alexander Denny, Dumbarton, and resold to Alexander A Laird; 1858 sold to Thomas Steele & David Hunter, Ayr; 1858 sold to Thomas Steele, David Hunter & James Paton, Ayr; 1858 passed to Ayr Steamship Co Ltd; 23 August 1859 stranded off Maughold Head, Isle of Man, and became a wreck.
Eagle	1857-1859	324	Built by Alexander Denny, Dumbarton, 1857, as *Eagle* for Alex Denny, Dumbarton, and sold to Alexander A Laird; 26 November 1859 run down and sank off Lamlash on a voyage Liverpool via Ramsey to Glasgow.
Falcon	1860-1867	390	Built by Archibald Denny, Dumbarton, 1860, as *Falcon* for Alexander A Laird (City of Glasgow Steam Packet Co); 1863 sold to McConnell & Laird; 1864 sold to Alexander A Laird; 8 January 1867 wrecked near Machrihanish, when seeking shelter of Mull of Kintyre (Glasgow for Londonderry) during snowstorm on voyage Glasgow to Londonderry.
Erin	1867		See *Rose* below.
Scotia	1867	208	See *Laurel* below.

McConnell & Laird also managed, but did not own, the paddle steamers belonging to Kemp & Company: *Royal Consort* (1844) between 1853 and 1873; *Princess Alice* (1843) between 1853 and 1858; *Fenella* (1846) between 1853 and 1854; *Cambria* (1845) between 1853 and 1854; *Nile* (1837) between 1853 and 1854.

Glasgow & Londonderry Steam Packet Company – Glasgow agent Alexander A Laird (1867-1873)

Name	G&LSP Co Service	Gross tons	Comments
Rose	1867-1883	409	Built by Blackwood & Gordon, Port Glasgow, 1867, as *Erin* for Alexander A Laird (MacConnell & Laird) completed for Glasgow & Londonderry Steam Packet Co and registered in names of Trustees; 1883 sold to Ayr Steam Shipping Co, Ayr, and registered in names of Trustees; 1889 renamed *Ailsa*; 26 February 1892 wrecked on Island Magee, Co. Antrim, on voyage from Belfast to Ayr.
Laurel	1867-1877	208	Built by Laurence Hill & Co, Port Glasgow, 1864, as *Jamaica Packet* for James & David Carson, Glasgow; 1867 sold to Alexander A Laird & Martin Orme (MacConnell & Laird) and renamed *Scotia*; 1867 sold to Glasgow & Londonderry Steam Packet Co and registered in names of Trustees, renamed *Laurel*; 1876 new engines; 1877 sold to James Duthie; 1877 sold to David Leckie & William Esplin (manager of Montrose, Arbroath, London & Newcastle Steamship Companies); 1880 sold to Watson, Kipling & Co Ltd; 1 October 1880 sank off Mouse lighthouse, Thames estuary, after a collision with the German vessel *Portia*, on voyage from Seaham to London.
Vine	1867	417	Built by Randolph, Elder & Co, Govan, Glasgow, 1867 as *Vine* for Glasgow & Londonderry Steam Packet Co but completed for D R McGregor & others, Leith; 1868 sold to Oriental Screw Collier Co; 1871 sold to Robert Mowbray; 1872 sold to Det Sondenfjelds Norske D/S (T Dannewig), Christiania, Norway, and renamed *Kong Sigurd*; 26 November 1902 abandoned sinking in Bay of Biscay on voyage from Swansea to Valencia.

Name	G&LSP Co Service	Gross tons	Comments
Garland	1868-1874	594	Paddle steamer built by Caird & Co, Greenock, 1863, as *Roe* for J & G Burns, but completed as *City of Petersburg* for George Campbell, Dunoon, and others, and used as blockade runner; 1865 sold to George Campbell, Liverpool; 1865 sold to Liverpool & Dublin Steam Navigation Co Ltd, Liverpool, and renamed *Bridgewater*; 1868 sold to Alexander A Laird, Glasgow, resold to Glasgow & Londonderry Steam Packet Co and renamed *Garland*; 1873 new engines; 1874 sold to Edward M de Bussche, Ryde, Isle of Wight; 1876 sold under mortgage to Thomas Redway, Exmouth; 1880 sold under mortgage to Warrenpoint Steam Packet Co, Warrenpoint; 1881 sold to Donald McCall, Greenwich; 1881 sold to James A Bayley, London; 1881 scrapped.
Fern	1873-1897	270	Built by McCullough, Patterson & Co, Port Glasgow, 1871 as *Gala* for Berwick & London Steam Ship Co Ltd, Berwick; 1873 sold to Glasgow & Londonderry Steam Packet Co. and renamed *Fern*; 1880 new engines; 1885 passed to Glasgow, Dublin & Londonderry Steam Packet Co Ltd (Alexander A Laird & Co); 1897 sold to Vicomte le Guales de Mezaubran, Le Légué, France and renamed *Le Légué*; 5 July 1990 struck rock and sank south east of Jersey, on voyage from St. Brieuc to Jersey.

Glasgow & Londonderry Steam Packet Company – Glasgow agent Alexander A Laird & Company (1873-1885)

Name	G&LSP Co Service	Gross tons	Comments
Holly	1875-1899	378	Built by D & W Henderson & Co Ltd, Partick, 1875, as *Holly* for Glasgow & Londonderry Steam Packet Co; 1885 passed to Glasgow, Dublin and Londonderry Steam Packet Co Ltd; 1899 sold to Joseph Madden; 1899 sold to A Diakiki & Co, Piraeus, Greece, and renamed *Piraeus*; 1909/10 sold to La Nav. Hellénique A Diakaki (Anastassios I Diakakis); 1912/13 sold to G Markettos and renamed *Halcyon*; 1914/15 sold to Navigation à Vapeur 'Ionienne' G Yannoulatou Frères; 1919/20 sold to National Steam Navigation Co Ltd of Greece (Embiricos Brothers) and renamed *Serifos*; 1922/23 sold to Steam Navigation of Samos (D Inglessi Fils), Samos, Greece; 1924-26 sold to D Inglesi Fils S A Nav. de Samos; 1935/36 scrapped.
Arbutus	1875-1878	637	Built by A & J Inglis, Pointhouse, 1875, as *Arbutus* for Glasgow & Londonderry Steam Packet Co; 18 April 1878 aground on rocks near Southend, Mull of Kintyre, on voyage from Londonderry to Glasgow, slid off rocks & sank.
Iris	1876-1883	625	Built by A & J Inglis, Pointhouse, 1876 as *Iris* for Glasgow & Londonderry Steam Packet Co; 2 September 1883 wrecked 1 mile east of Inishtrahull lighthouse, on voyage from Glasgow to Sligo.
Vine	1878-1881	488	Built by D & W Henderson & Co Ltd, Partick, 1878, as *Vine* for Alexander A Laird and passed to Glasgow & Londonderry Steam Packet Co (Alexander A Laird & Co); 1881 sold to Constantine C Ralli, Liverpool; 1882 sold to Panhellenic Steam Ship Navigation, Piraeus, Greece, and renamed *Argolis*; 1896 passed to Navigation à Vapeur Panhellénique, Piraeus, Greece; 1907/08 sold to Cie de Navigation à Vapeur 'Ermopolis' (E A Foustanos, Syra, Greece; 1918/19 sold to Hellenic Co of Maritime Enterprises (A Palios); 1924/25 sold to J Lykouris and renamed *Maria L*; 1940 renamed *Canisbay* and renamed *Milos*, registered in Panama; 1941 sold to Palestine Maritime Lloyd Ltd, Haifa, Palestine, and renamed *Atzila*; 4 November 1941 sank at Port Said.

Name	Years	No.	Details
Azalea	1878-1914	706	Built by A & J Inglis, Pointhouse, 1878, as *Azalea* for Glasgow & Londonderry Steam Packet Co (Alexander A Laird & Co); 1885 passed to Glasgow, Dublin & Londonderry Steam Packet Co Ltd (Alexander A Laird & Co); 1907 passed to Laird Line Ltd; 1914 sold to Navigation à Vapeur 'Ionienne' G Yannoulatou Frères, Piraeus, Greece, and renamed *Chalkis* and resold to G Gambetta; 1917 renamed *Nafkratousa*; 1920 sold to Paneveiki; 1922 sold to Navigation à Vapeur 'Ionienne' G Yannoulatou Frères; 1925 sold to S A Ionienne de Navigation à Vapeur "Yannoulatos", Piraeus, Greece; 1927 sold to Coast Line Steamship Co of Greece; 1929 passed to Hellenic Coast Lines Co Ltd, Piraeus, Greece, and renamed *Nafcratousa*; 1933 renamed *Psara*; 1939 scrapped.
Cedar	1878-1906 1908-1924	719	Built by D & W Henderson & Co Ltd, Meadowside, 1878, as *Cedar* for Alexander A Laird & Co; 1885 to Glasgow, Dublin & Londonderry Steam Packet Co Ltd (Alexander A Laird & Co); 1894 passed to Alexander A Laird & Co; 1906 sold to Ayr Steam Shipping Co Ltd and renamed *Dunure*; 1908 Ayr Steam Shipping Co Ltd acquired by Laird Line Ltd. 1921 to Laird Line Ltd; 1922 to Burns & Laird Lines Ltd; 1924 sold to Ph Kavounides, Piraeus, Greece, and renamed *Nicolaos Kavounides*; 1926 renamed *Bosphoros*; 1927 renamed *Express*; 1928 sold to Alex A Yannoulatos, Piraeus, Greece, and renamed *Zephiros*; 1929 Alex A Yannoulatos acquired by Hellenic Coast Lines Co Ltd; 1933 renamed *Spetsai*; 1937 scrapped.
Shamrock	1879-1918	832	Built by A & J Inglis, Pointhouse, 1879, as *Shamrock* for Glasgow & Londonderry Steam Packet Co (Alexander A Laird & Co) ; 1885 to Glasgow, Dublin & Londonderry Steam Packet Co Ltd (Alexander A Laird & Co); 1891 to A A Laird & Co; 5 May 1918 aground at Freshwater Bay, Lambay Island, Co. Dublin; sold in damaged condition to Hammond Lane Foundry Co and demolished.
Brier	1882-1931	710	Built by D & W Henderson & Co Ltd, Meadowside,1882, as *Brier* for Alexander A Laird; 1883 to Glasgow & Londonderry Steam Packet Co; 1885 to Glasgow, Dublin & Londonderry Steam Packet Co Ltd (Alexander A Laird & Co); 1891 to Alexander A Laird & Co; 1898 to Glasgow, Dublin & Londonderry Steam Packet Co Ltd; 1907 to Laird Line Ltd; 1919 Laird Line Ltd acquired by Coast Lines Ltd; 1922 to Burns & Laird Lines Ltd; 1929 renamed *Lairdsoak*; 1931 sold to Michael Murphy Ltd, Dublin, and renamed *Enda*; 27 February 1933 wrecked in Clanyard Bay, Mull of Galloway, on voyage from Londonderry to Heysham while on charter to Burns & Laird Lines Ltd.
Rose	1883-1884	449	Built by A Stephen & Sons, Linthouse, 1883, as *Daphne*; sank on launching; 1883 completed for Glasgow & Londonderry Steam Packet Co (Alexander A Laird & Co) as *Rose*; 1 March 1844 aground on Farlane Point, Millport, in fog. Refloated, sold to John Bell, Prestwick. Rebuilt & renamed *Ianthe*. New engines; 1889 sold by executors of John Bell (died 1887) to Henry J Handcock; 1889 sold to Navigation Orientale; 1891 sold to P Pantaleon, Smyrna, Greece, and renamed *Eleni*; 1896 to Navigation Orientale (P Pantaleon) Smyrna, Greece; 10 December 1918 mined and sank off Tenedos.
Daisy	1884-1889	225	Built by H McIntyre & Co, Paisley, 1884, as *Daisy* for Alexander A Laird & Co; 1885 to Glasgow, Dublin & Londonderry Steam Packet Co Ltd (Alexander A Laird & Co); 1889 sold to Aktieselskabet J Mortensens Efterfolgere, Trangisvaag, Faroe Islands, Denmark, and renamed *Föringur*; 1900 sold to George Couper, Helmsdale, Sutherland, and renamed *Daisy*; 1900 sold to Peterhead, Aberdeen & Leith Steam Navigation Co Ltd; 13 March 1901 wrecked at Cruden Skares, Bay of Cruden, Peterhead, on voyage from Aberdeen to Peterhead.

Name			Comments
Thistle	1884-1928	803	Built by D & W Henderson & Co Ltd, Meadowside,1884, as *Thistle* for Alexander A Laird; 1884 to Alexander A Laird & Co; 1885 to Glasgow, Dublin & Londonderry Steam Packet Co Ltd (Alexander A Laird & Co); 1891 to Alexander A Laird & Co; 1903 lengthened by 15 feet, and to Glasgow, Dublin & Londonderry Steam Packet Co Ltd (Alexander A Laird & Co); 1907 to Laird Line Ltd; 1919 Laird Line Ltd acquired by Coast Lines Ltd; 1922 to Burns & Laird Lines Ltd; 1928 sold to Smith & Co to be broken up at Port Glasgow; 1929 scrapped.
Elm	1884-1910	489	Built by A & J Inglis, Pointhouse, 1884, as *Elm* for Alexander A Laird; 1884 to Alexander A Laird & Co; 1885 to Glasgow, Dublin & Londonderry Steam Packet Co Ltd (Alexander A Laird & Co); 1891 to Alexander A Laird & Co; 1898 to Glasgow, Dublin & Londonderry Steam Packet Co Ltd; 1907 to Laird Line Ltd; 1910 sold to Walter Couper; 1910 sold to Empreza Navigazion Bahiana, Bahia, Brazil, and renamed *Guararapes*; 1916/17 deleted from Lloyd's Register.

Glasgow, Dublin & Londonderry Steam Packet Company Limited – agent Alexander A Laird & Company (1885-1907)

Name	GDLSPC Service	Gross tons	Comments
Gardenia	1885-1904	461	Built by D & W Henderson & Co Ltd, Meadowside, 1885, as *Gardenia* for Alexander A Laird & Co and passed to Glasgow, Dublin & Londonderry Steam Packet Co Ltd (Alexander A Laird & Co); 1904 sold to Transports Maritimes Est Tunisien (P Normant), Le Havre, France; 7 November 1909 foundered after springing a leak off St. Tropez, on voyage from Marseilles to Nice.
Ivy	1888-1900	531	Built by C J Bigger, Londonderry, 1888, as *Ivy* for Glasgow, Dublin & Londonderry Steam Packet Co Ltd (Alexander A Laird & Co); 1891 to Alexander A Laird & Co; 1900 sold to D Bloom & Co, San Francisco, USA; 1903 sold to Colombian Government and renamed *Padilla*; 1905 sold to J J McAuliffe, Coquimbo, Chile, and renamed *Ivy*; 1906 sold to Antofagasta & Bolivia Railway Co, Chile; 1908 to Antofagasta, Chile & Bolivia Railway Co; 1913 sold to F G Piaggio & Co, Callao, Peru, and renamed *Fenice*; 9 April 1921 wrecked at Lobos de Tierra Islands.
Stanley	1890	715	Paddle steamer built by Caird & Co, Greenock, 1864, as *Stanley* for London & North Western Railway Co; 1888 sold to Irish National Steamship Co Ltd, Lough Derg; 1890 sold to Glasgow, Dublin & Londonderry Steam Packet Co Ltd; 1890 scrapped at Bowling.
Laurel	1890-1891	1000	Paddle steamer built by A Leslie & Co, Hebburn, Newcastle, 1873, as *Thomas Dugdale* for Lancashire & Yorkshire and London & North Western Railway Companies; 1882 new engines; 1888 sold to Irish National Steamship Co Ltd, Londonderry; 1890 sold to Glasgow, Dublin & Londonderry Steam Packet Co Ltd and renamed *Laurel*; 1891 sold to William R. Fairlie and scrapped.
Olive	1893-1930	1141	Built by D & W Henderson & Co Ltd, Meadowside, Glasgow, 1893, as *Olive* for Glasgow, Dublin & Londonderry Steam Packet Co Ltd (Alexander A Laird & Co); 1907 to Laird Line Ltd; 1919 Laird Line Ltd acquired by Coast Lines Ltd; 1922 to Burns & Laird Lines Ltd; 1929 renamed *Lairdsbank*; 1930 sold to North of Scotland & Orkney & Shetland Steam Navigation Co Ltd, Aberdeen, and renamed *St Catherine*; 1937 scrapped at Rosyth.
Daisy	1895-1915	565	Built by Blackwood & Gordon, Port Glasgow, 1895, as *Daisy* for Glasgow, Dublin & Londonderry Steam Packet Co Ltd (Alexander A Laird & Co); 1907 to Laird Line Ltd; wrecked 23 February 1915 at the bar of the River Bann, on voyage from Greenock to Coleraine.

Name	Laird service	Gross tons	Comments
Lily	1896-1937	668	Built by Blackwood & Gordon, Port Glasgow, 1896, as *Lily* for Glasgow, Dublin & Londonderry Steam Packet Co Ltd (Alexander A Laird & Co); 1907 to Laird Line Ltd; 1919 Laird Line Ltd acquired by Coast Lines Ltd; 1922 to Burns & Laird Lines Ltd; 1929 renamed *Lairdspool*; 1937 to David MacBrayne Ltd and renamed *Lochgorm*; 1951 scrapped at Port Glasgow.
Fern	1900-1918	503	Built by Ailsa Shipbuilding Co Ltd, Troon, 1900, as *Fern* for Glasgow, Dublin & Londonderry Steam Packet Co Ltd (Alexander A Laird & Co); 1907 to Laird Line Ltd; 22 April 1918 torpedoed by U104 and sank 10 miles east of the Baily light, Dublin Bay, on voyage from Dublin to Heysham.
Rose	1902-1949	1151	Built by A & J Inglis Ltd, Pointhouse, 1902, as *Rose* for Glasgow, Dublin & Londonderry Steam Packet Co Ltd (Alexander A Laird & Co); 1907 to Laird Line Ltd; 1919 Laird Line Ltd acquired by Coast Lines Ltd; 1922 to Burns & Laird Lines Ltd; 1929 renamed *Lairdsrose*; 1949 scrapped.

Laird Line Limited (1907-1922)

Name	Laird service	Gross tons	Comments
Hazel	1907-1919	1241	Built by Fairfield Shipbuilding & Engineering Co Ltd, Govan, 1907, as *Hazel* for Laird Line Ltd (Alexander A Laird & Co); 1914 requisitioned as fleet messenger; 1919 returned to owners; 1919 sold to Isle of Man Steam Packet Co Ltd, Douglas, and renamed *Mona*; 1938 scrapped.
Turnberry	1908-1937	565	Built by Archibald McMillan & Sons, Dumbarton, 1889, as *Spindrift* for C F Leach, London; 1899 sold to Ayr Steam Shipping Co Ltd, Glasgow, and renamed *Turnberry*; 1908 Ayr Steam Shipping Co Ltd acquired by Laird Line Ltd; 1919 Laird Line Ltd acquired by Coast Lines Ltd; 1922 to Burns & Laird Lines Ltd; 1929 renamed *Lairdsheather*; 1937 scrapped.
Dunure	1908-1924	719	See *Cedar* page 114.
Maple	1914-1951	1304	Built by Ailsa Shipbuilding Co Ltd, Troon, 1914, as *Maple* for Laird Line Ltd (Alexander A Laird & Co); 1919 Laird Line Ltd acquired by Coast Lines Ltd; 1922 to Burns & Laird Lines Ltd; 1929 renamed *Lairdsglen*; 1951 scrapped at Port Glasgow.
Broom	1915-1922	626	Built by Ramage & Ferguson Ltd, Leith, 1904, as *James Crombie* for Aberdeen, Leith & Moray Firth Steam Shipping Co Ltd, Aberdeen; 1914 Aberdeen, Leith & Moray Firth Steam Shipping Co Ltd acquired by M Langlands & Sons; 1915 sold to Laird Line Ltd (Alexander A Laird & Co) and renamed *Broom*; 1919 Laird Line Ltd acquired by Coast Lines Ltd; 1922 to City of Cork Steam Packet Co Ltd, Cork, and renamed *Lismore*; to Belfast Steamship Co Ltd, Belfast, and renamed *Dynamic*; 1932 renamed *Ulster Star*; 1940 chartered to David MacBrayne Ltd; 1949 scrapped.
Cairnsmore	1921-1929	877	Built by Harland & Wolff Ltd, Belfast, 1896, as *Comic* for Belfast Steamship Co Ltd, Belfast; 1921 to Laird Line Ltd and renamed *Cairnsmore*; 1922 to Burns & Laird Lines Ltd; 1929 to British & Irish Steam Packet Co Ltd, Dublin, and renamed *Lady Kerry*; 1934 scrapped.
Culzean	1921-1929	883	Built by Barclay, Curle & Co Ltd, Whiteinch, 1897, as *Logic* for Thomas Gallaher, Belfast (Belfast Steamship Co Ltd); 1904 to Belfast Steamship Co Ltd, Belfast; 1921 sold to Laird Line Ltd and renamed *Culzean*; 1922 to Burns & Laird Lines Ltd; 1929 to British & Irish Steam Packet Co Ltd, Dublin, and renamed *Lady Carlow*; 1936 sold to Smith & Co for breaking up; 1937 resold and scrapped.

Ayr Steam Shipping Company Limited (owned by Laird Line Ltd from 1908 merged into Laird Line 1921)

Name	Ayr SSC service	Gross tons	Comments
Turnberry	1899-1922	565	See *Turnberry* page 116.
Merrick	1903-1908	586	Built by Birrell, Stenhouse & Co, Dumbarton, 1878, as *Amsterdam* for James Rankine, Glasgow; 1903 sold to Ayr Steam Shipping Co Ltd, Rowan & Bain, Ayr, and renamed *Merrick*; 29 December 1908 wrecked near Lendalfoot, 6 miles south west of Girvan, on voyage Ayr to Larne.
Dunure	1906-1922	719	See *Cedar* page 114.

See *Clyde and Other Coastal Steamers*, Duckworth and Langmuir, for details of steamers owned by Ayr Steam Shipping Co before May 1908.

2) Burns and associate companies

James Martin & J & G Burns, J & G Burns (1824 – 1842) – Irish services

Name	J&G Burns Service	Gross tons	Comments
Fingal	1826-1835	350	Paddle steamer built by William Simons & Co, Greenock, 1826, as *Fingal* for Belfast and Glasgow Steam Boat Co, registered at Belfast; 1835 sold to Robert Purdon, Newry; 1838 sold to George McTear, Belfast; 1846 vessel declared unseaworthy
Eclipse	1828-1837	174	Paddle steamer built by James Lang, Dumbarton, 1826, as *Eclipse* for James Dalzell, Robert Napier, James Stevenson, David Napier & James Moore, Glasgow; 1827 chartered to Lewis MacLellan & others for Glasgow - Londonderry service; 1828 sold to James Dalzell, Robert Napier, James Stevenson & David Napier, Glasgow; 1828: Sold to Robert Napier, James Dalzell, James Stevenson & David Napier, Glasgow; 1828 chartered to J Martin and J & G Burns for Glasgow - Belfast service; 1828 sold to Belfast and Glasgow Steam Boat Co, Belfast; 1837 sold to James Leacy, Liverpool; 1838 sold to Francis Scallan & James Leacy, Liverpool; 1839 sold to Marks Doyle, Liverpool; 1839 sold to James Leacy & John Tarleton, Liverpool; 1840 sold to John Jones, Carnarvonshire, James Leacy & John Tarleton, Liverpool; 1841 sold to William Murphy, Newtown, Co. Wexford, John Jones, Carnarvonshire, and John Tarleton; 1841 sold to Thomas McTear & William Murphy, Newtown, Co. Wexford; 1856 scrapped.
Belfast	1830-1837	123 net	Paddle steamer built by John Wood & Joseph H Ritchie, Port Glasgow, 1829, as *Belfast* for David Napier, Glasgow, and chartered to John Gemmill; 1830 sold to Belfast & Glasgow Steam Boat Co, Belfast; 1835 sold to Belfast & Glasgow Steam Shipping Co, Belfast; 1837 lengthened 13 feet; 1837 sold to the Admiralty and renamed *Prospero* for use as packet on Milford - Waterford service; 1838 mercantile register closed; 1853 converted to passenger tender; 1866 scrapped.
Antelope	1833-1845	162 net	Paddle steamer built by Robert Barclay, Glasgow, 1833, as *Antelope* for Belfast and Glasgow Steam Boat Co, Belfast; 1838 to Belfast and Glasgow Steam Shipping Co, Belfast; 1840 sold to James Burns & George Burns (James & George Burns), Glasgow; 1844 to Glasgow & Liverpool Steam Shipping Co; 1845 sold to William B Brownlow, Hull; 1848 sold to William B Brownlow and others (Hull Steam Packet Co), Hull; 1856 scrapped.
Rapid	1835-1838	402 net	Paddle steamer built by Hunter and Dow, Glasgow, 1835, as *Rapid* for Belfast and Glasgow Steam Shipping Co, Belfast; 1838 sold to Berwick Shipping Co, Berwick; 1845 sold to Anthony G Robinson, London; 1845 sold to Anthony G Robinson and Thomas Pim; 1846 sold to Malcomson Brothers, Waterford; 1852 scrapped.

Name	Service	Tons	Comments
Mercury	1836-1837	200 net	Paddle steamer built by J Wood, Port Glasgow, 1835, as **Mercury** for William Coates and J Young, Belfast; 1836 sold to Belfast & Glasgow Steam Shipping Co, Belfast; 1837 sold to Dublin & Glasgow Sailing & Steam Packet Co, Dublin; 1850 wrecked.
Aurora	1839-1849	453	Paddle steamer built by Charles Connell & Sons, Belfast, 1839, as **Aurora** for J & G Burns, Glasgow; 1844 to Glasgow and Liverpool Steam Shipping Co; 1850 converted to sailing ship, barque rigged; 7 October 1850 foundered off Rhoscolyn, Anglesey on voyage from Cardiff to Dublin.
Fire King	1842-1846	526	Paddle steamer built by Robert Thomson, Troon, 1838, as yacht **Fire King** for T A Smith, Beaumaris; 1840 sold to Robert Napier, Glasgow; 1840 sold to James McCall, John Leadbetter, William Brooks, Archibald Smith and Archibald Harvie; 1842 sold to James & George Burns, Glasgow; 1844 sold to Glasgow & Liverpool Steam Shipping Co; 1846 sold to Alexander Bell, London; 1847 sold overseas and register closed.

J Martin & J & G Burns (1824 - 1842) – Liverpool services, traded as Glasgow and Liverpool Steam Shipping Co.

Name	JM & J&G B Service	Gross tons	Comments
Glasgow	1829-1837	181 net	Paddle steamer built by John Wood & Joseph H Ritchie, Port Glasgow, 1829, as **Glasgow** for Glasgow and Liverpool Steam Shipping Co; 1834 lengthened by 18 feet; 1837 sold to John Townley, Thomas Coleman, Charles Duffy, James Carrol, Nicholas Martin, Patrick Wynne, Neal Kelly and John Gairtlaire, Dundalk; 1837 sold to Dundalk Steam Packet Co; 1845 scrapped.
Ailsa Craig	1829-1838	110 net	Paddle steamer built by R & A Carswell, Greenock, 1825, as **Ailsa Craig** for Ailsa Craig Steam Yacht Co, Glasgow; 1827 sold to Robert Steele & Co, Agnew Crawford and Matthew Brownlie, Glasgow; 1829 sold to Robert Steele, junior, & Matthew Brownlie, lengthened by 30 feet, resold to Glasgow and Liverpool Steam Shipping Co; 1838 sold to Daniel Harmer, Norwich and resold to Norfolk Steam Packet Co; 1846 scrapped.
Liverpool	1830-1835	196 net	Paddle steamer built by Robert Steele & Co, Greenock, 1830, as **Liverpool** for Glasgow and Liverpool Steam Shipping Co; 1835 sold to Richard Bourne and others, Dublin; 1841 new engines; 1842 sold to Peninsular and Oriental Steam Navigation Co, London; November 1845 stranded at Tarifa, Spain, refloated, but condemned and broken up.
Henry Bell	1831	113 net	Paddle steamer built by Wilson & Gladstone, Liverpool, 1823, as **Henry Bell** for Mersey and Clyde Steam Navigation Co, Liverpool; 1831 Mersey and Clyde Steam Navigation Co acquired by Glasgow and Liverpool Steam Shipping Co; 1831 sold to John Scott & Robert Sinclair, Greenock, and resold to Robert Purdon, Newry; 1840 scrapped.
James Watt	1831	116 net	Paddle steamer built by Humble & Hurry, Liverpool, 1824, as **James Watt** for Mersey and Clyde Steam Navigation Co, Liverpool; 1829 lengthened by 15 feet; 1831 Mersey and Clyde Steam Navigation Co acquired by Glasgow and Liverpool Steam Shipping Co; 1831 sold to New Merchants Shipping Co, of Stockton; 1836 sold to Joseph H Grose, Sydney, Australia; 1836 sold to Thomas Street and Joseph H Grose, Sydney, Australia (Grose & Street) and converted to sailing ship, schooner rigged; 1839 sold to Joseph H Grose, Sydney, Australia; 1840 re-converted to steamship; 1842 sold to Robert Scott, John Hocking and Henry T Keldon, Sydney, Australia; 1846 sold to Hunter River Steam Navigation Co; 1847 scrapped.

William Huskisson	1831-1832	213 net	Paddle steamer built by John Scott and Sons, Greenock, 1826, as *William Huskisson* for Mersey and Clyde Steam Navigation Co, Liverpool; 1831 Mersey and Clyde Steam Navigation Co acquired by Glasgow and Liverpool Steam Shipping Co; 1832 sold to John Watson and managed by Saint George Steam Packet Co; 1832 sold to City of Dublin Steam Packet Co; 12 January 1840 abandoned off Holyhead, on voyage from Dublin to Liverpool after springing a leak. Later sank in approaches to River Mersey.
Enterprize	1831	125 net	Paddle steamer built by William Denny, Dumbarton, 1826, as *Enterprize* for New Clyde Shipping Co, Glasgow; 1831 sold to Glasgow and Liverpool Steam Shipping Co, Glasgow, and resold to Stockton Shipping Co; 1833 sold to Thomas Hudson, William Cobby & Robert J Benyon, Hull; 1834 sold to Robert J Benyon and others, London, and lengthened by 13 feet; 1835 sold to William Cobby and others; 1835 sold to Humber Union Steam Co, Hull; 1843 sold to T Wise, Boston; 1844 sold to T Wise and C J Griffin, London; 1844 sold to T Wise, C J Griffin and others; 1846 sold to W Briggs, W Johnson, R Ferrier and others; 1853 sold to Thomas H Holderness, Liverpool, and Mark Blower, Yarmouth; 1859 scrapped.
Clyde	1831-1837	196 net	Paddle steamer built by MacMillan & Duncan, Greenock, 1831, as *Clyde* for Glasgow and Liverpool Steam Shipping Co; 1837 sold to London & Havre Steam Packet Co, London; 1839 sold to T C Gibson, London; 1839 sold to C E Hoppe, London; 1839 sold to Newcastle, Hamburg & Rotterdam Steamship Co, Newcastle; 1840 sold to Thomas C Gibson, Newcastle; 1841 register closed following a fire at Corunna.
Manchester	1832-1838	412 net	Paddle steamer built by Robert Steele & Co, Greenock, 1832, as *Manchester* for Glasgow and Liverpool Steam Shipping Co, Glasgow; 1838 sold to Berwick Shipping Co, Berwick; 1840 sold to Joseph Somes, Middlesex; 1847 sold to Berwick Shipping Co, Berwick; 1855 register closed (coal hulk).
Antelope	1833-1845	162 net	See page 117.
Gazelle	1833-1834	187 net	Paddle steamer built by Murries & Clark, Greenock, 1832, as *Gazelle* for John Fleming, Glasgow; 1833 sold to Glasgow and Liverpool Steam Shipping Co, Glasgow; 1834 sold to William B Brownlow, William H Pearson, John Holmes, Charles Darley and others (Hull Steam Packet Co), Hull; 1844 lengthened by 16 feet; 1863 sold to J D Schultz, Hamburg, Germany; 1865 converted to dumb barge.
Eagle	1835-1839	293 net	Paddle steamer built by Robert Steele & Co, Greenock, 1835, as *Eagle* for Glasgow and Liverpool Steam Shipping Co, Glasgow; 1839 sold to Dublin & Glasgow Steam Packet Co, Dublin; 1846 sold to Royal Mail Steam Packet Co, London; 1849 re-built; 1859 laid up in Jamaica; 1860 register closed.
Unicorn	1836-1840	648	Paddle steamer built by Robert Steele & Co, Greenock, 1836, as *Unicorn* for Glasgow and Liverpool Steam Shipping Co, Glasgow; 1840 sold to British & North American Royal Mail Steam Packet Co, Glasgow; 1845 sold to James Whitney, Saint John, New Brunswick; 1846 used as corvette in Portuguese Navy; 1849 sold to Samuel Cunard and resold to Edward Cunard, New York, USA; resold to Pacific Mail Steamship Co, San Francisco, USA; 1854 sold to Edye Manning, Sydney, Australia; 1856 sold to Thomas Hunt & Co; 1857 renamed *E H Green*; 1858 laid up at Whampoa, Hong Kong; 1872 register closed.
Actæon	1837-1841	684	Paddle steamer built by Robert Steele & Co, Greenock, 1837, as *Actæon* for Glasgow and Liverpool Steam Shipping Co; 1841 sold to Royal Mail Steam Packet Co, London; 9 October 1844 wrecked on Negrel Rock, near Cartagena, Colombia, on voyage from Santa Marta to Cartagena.
Aurora	1839-1849	453	See page 118.

Name	Service	Tons	Comments
Achilles	1839-1845	889	Paddle steamer built by Robert Steele & Co, Greenock, 1839, as *Achilles* for Glasgow and Liverpool Steam Shipping Co, Glasgow; 1840 lengthened by 23 feet; 1845 sold to Peninsular & Oriental Steam Navigation Co, London; 1856 scrapped.
Fire King	1842-1846	526	See page 118.
Foyle	1844-1847	136	See page 108.

William Young & George Burns, George Burns and Glasgow & Liverpool Steam Shipping Company, J Martin & J & G Burns 1835 -1851 – Clyde and West Highland services

Name	JM & J&G B Service	Gross tons	Comments
Inverness	1835-1846	43 net	Paddle steamer built by Robert Barclay, Glasgow, 1832, as *Inverness* for George Smith, James Melvin, Peter Turner and William Young, Glasgow; 1835 sold to William Young and resold to William Young and George Burns and resold again in September to George Burns; re-engined; 1844 sold to Glasgow & Liverpool Steam Shipping Co; 1846 sold to Samuel Bromhead & Samuel Hemming (Bromhead & Hemming), Londonderry; 1848 scrapped.
Rob Roy	1835-1849	42 net	Paddle steamer built by R Duncan & Co, Greenock, 1834, as *Rob Roy* for William Young, Glasgow; 1835 sold to William Young & George Burns; re-engined and resold in September to George Burns; 1844 sold to Glasgow & Liverpool Steam Shipping Co; 1850 scrapped.
Helen McGregor	1835-1848	50 net	Paddle steamer built by R. Duncan & Co, Greenock, 1835, as *Helen McGregor* for George Burns and William Young, Glasgow; 1835 sold to George Burns; 1844 re-built and sold to Glasgow & Liverpool Steam Shipping Co, Glasgow; 1849 scrapped (laid up in 1848).
Glenalbyn	1837-1838	200 net	Paddle steamer built by John Scott & Co, Greenock, 1834, as *Glen Albyn* for Glen Albyn Steamboat Co, Tobermory; 1835 sold to James J Duncan, West of Scotland Insurance Co, Glasgow, and renamed *Glenalbyn*; 1837 sold to J Martin & J & G Burns (North British Steam Navigation Co); 1838 sold to General Shipping Co,, Berwick-on-Tweed; 1841 sold to Hull & Leith Steam Packet Co, Hull; 1852 sold to Leith, Hull and Hamburgh Steam Packet Co, Leith; 1853 sold to T, C L & J N Ringrose, Hull; 3 March 1856 wrecked on West Plaat, near Brielle, off the entrance to River Maas on voyage Grimsby to Rotterdam.
Dolphin	1844-1851	248	Paddle steamer built by Robert Napier, Govan, 1844, as *Dolphin* for Scottish Central Railway Co, but completed for Glasgow & Liverpool Steam Shipping Co; 1851 ceded to David Hutcheson & Co, Glasgow; 1862 sold to William J R Grazebrook, Liverpool, as blockade runner; 1863 lengthened by 9 feet and then captured by U.S.S. *Wachusett*; 1864 auctioned at New York and renamed *Annie*; 1867 sold to owners in Memphis, Tennessee; 1874 wrecked.
Culloden	1844-1851	149	Paddle steamer built by Caird & Co, Greenock, 1844, as *Culloden* for Glasgow & Liverpool Steam Shipping Co; 1851 sold to William Denny & Brothers, Dumbarton; 1851 sold to Henry P Maples and William Denny & Brothers, London, for Newhaven - Dieppe service, and resold to William Denny & Brothers, London, on completion of that service; 1852 sold to James Denny, Dougald MacDougall, William White, Peter Roger William Swan Stuart and William Denny & Brothers; 1853 sailed from Glasgow for Melbourne as schooner; 1853 sold to Derwent & Huon Steam Navigation Co, Hobart Town, engines re-installed; 1865 sold to Gourlay & Armstrong, Hobart Town; 1866 sold to Edwin T Beilby, Sydney, Australia; 1866 sold to R Whitaker & John Broomfield, Sydney, Australia; 1870 to John Broomfield, Sydney, Australia; 1871 sold to Matthew Johnston, Sydney, Australia; 28 April 1872 wrecked on Richmond river bar.

Maid of Islay	1846	91 net	Paddle steamer built by John Wood & Co, Port Glasgow, 1824, as *Maid of Islay* for William F Campbell, Islay (also known as *Maid of Islay No. 2*); 1845 sold to Glasgow Castle Steam Packet Co, Glasgow; 1846 Glasgow Castle Steam Packet Co acquired by Glasgow and Liverpool Steam Shipping Co and resold to William C Townley; 25 May 1848 seized by HMS *Alert* in Sierra Leone, ordered by Court to be returned to owner, but no evidence of further trading.
Dunoon Castle	1846-1851	175	Paddle steamer built by William Denny, Dumbarton, 1826, as *Dunoon Castle* for Dunoon Castle Steam Packet Co; 1831 sold to James Ewing, Glasgow; 1832 sold to Castle Steam Packet Co, Glasgow; 1842 to Glasgow Castle Steam Packet Co, Glasgow; 1846 Glasgow Castle Steam Packet Co acquired by Glasgow and Liverpool Steam Shipping Co; 1851 sold to William Denny & Brothers, Dumbarton; 1851 sold to Thomas Brownlee (Glasgow & Lochfyne Steam Packet Co), Glasgow; 1854 sold to Michael McLachlan; 1854 to Catherine McLachlan following death of Michael McLachlan; 1855 sold to John McArthur; 1855 sold to Archibald McMeikan, Glasgow; 1856 reported scrapped.
Rothsay Castle	1846-1851	180	Paddle steamer built by Tod & McGregor, Glasgow, 1837, as *Rothsay Castle* for Castle Steam Packet Co, Glasgow; 1842 to Glasgow Castle Steam Packet Co, Glasgow; 1846 Glasgow Castle Steam Packet Co acquired by Glasgow and Liverpool Steam Shipping Co; 1851 sold to William Denny & Brothers, Dumbarton; 1851 sold to Henry P Maples & William Denny & Brothers, Dumbarton, for Newhaven - Dieppe service; 1851 sold to William Denny & Brothers, Dumbarton; 1853 sent to Australia under sail; 1857 register closed.
Inverary Castle	1846-1851	210	Paddle steamer built by Tod & McGregor, Glasgow, 1839, as *Inverary Castle* for Castle Steam Packet Co; 1842 sold to Glasgow Castle Steam Packet Co, Glasgow; 1846 Glasgow Castle Steam Packet Co acquired by Glasgow and Liverpool Steam Shipping Co; 1851 sold to William Denny & Brothers, Dumbarton; 1851 sold to Thomas Brownlie (Glasgow & Lochfyne Steam Packet Co); 1857 sold to David Hutcheson & Co, Glasgow; 1862 lengthened by 18 feet and renamed *Inveraray Castle*; 1873 lengthened by 14 feet; 1879 to David MacBrayne, Glasgow; 1892 scrapped.
Duntroon Castle	1846-1851	247	Paddle steamer built by Anderson & Gilmour, Govan, Glasgow, 1842, as *Duntroon Castle* for Glasgow Castle Steam Packet Co, Glasgow; 1846 Glasgow Castle Steam Shipping Co acquired by Glasgow and Liverpool Steam Shipping Co; 1851 ceded to David Hutcheson & Co, Glasgow; 1853 sold to Henry P Maples & Co for Newhaven - Dieppe service; 1854 sold to Henry P Maples; 1855/56 sold to A W Chalmers & Co; 1857 sold to Thomas Chilton, jnr., and others, London; 1863 sold to Ernest Drouke, Cork; 1 August 1863 sailed from Queenstown for Nassau, presumed foundered.
Cardiff Castle	1046-1851	206	Paddle steamer built by Caird & Co, Greenock, 1844, as *Cardiff Castle* for Glasgow Castle Steam Packet Co, Glasgow; 1846 Glasgow Castle Steam Packet Co acquired by Glasgow and Liverpool Steam Shipping Co; 1851 sold to William Denny & Brothers, Dumbarton; 1851 sold to Neil McGill & Dugald Weir; 1851 sold to Neil McGill, Dugald Weir & William F Johnstone; 1855 sold to John Ballardie & others; 1858 sold to William Buchanan & others; 1859 sold to Alexander Watson & Henry Sharp; 1866 sold to John Dore; 1867 scrapped.

Name	Years	No.	Details
Craignish Castle	1846-1851	206	Paddle steamer built by Caird & Co, Greenock, 1844, as **Craignish Castle** for Glasgow Castle Steam Packet Co, Glasgow; 1846 Glasgow Castle Steam Packet Co acquired by Glasgow and Liverpool Steam Shipping Co; 1851 sold to William Denny & Brothers, Dumbarton; 1851 sold to John Ballardie, John Reid, John Orme and John Shearer, Glasgow; 1854 sold to Archibald McGill; 1854 sold to Archibald McGill, Peter McGregor and John Workman; 1855 sold to Peter McGregor; 1863 sold to William McHutchison, Glasgow, following the death of Peter McGregor, re-sold to John Murphy, Birkenhead, re-sold to William J Grazebrook, Liverpool, and renamed **Adler**, intended as blockade runner; resold to James H Wilson, Liverpool and renamed **Triton**; 28 December 1863 put into Surinam damaged and severely strained, on voyage from Liverpool to Havana and condemned; 1864 repaired, sold to Gerard Stubbs, Georgetown; 1865 sold overseas.
Edinburgh Castle	1846-1851	114	Paddle steamer built by Smith & Rodger, Govan, 1844, as **Edinburgh Castle** for Glasgow Castle Steam Packet Co, Glasgow; 1846 Glasgow Castle Steam Packet Co acquired by Glasgow and Liverpool Steam Shipping Co; 1851 ceded to David Hutcheson & Co, Glasgow; 1860 sold to Robert Curle & James Hamilton (shipbuilders), Glasgow, and re-sold to David Hutcheson & Co, Glasgow; 1875 lengthened by 7 feet and renamed **Glengarry**; 1879 to David MacBrayne, Glasgow; 1902 to David MacBrayne, David Hope MacBrayne & Lawrence MacBrayne, Glasgow; 1905 to David MacBrayne & David Hope MacBrayne, Glasgow; 1906 to David MacBrayne Ltd & David MacBrayne, Glasgow; 1907 to David MacBrayne Ltd, Glasgow, following the death of David MacBrayne; 1927 scrapped.
Dunrobin Castle	1846-1851	207	Paddle steamer built by Tod & McGregor, Glasgow, 1846, as **Dunrobin Castle** for Glasgow and Liverpool Steam Shipping Co; 1851 sold to William Denny, James Denny and Peter Denny, junior (William Denny & Brothers), Dumbarton; 1851 sold to Russian owners and renamed **Telegraf**; 1853 sold to F Baird, Kronstadt, Russia; 1863 sold to Peterburgo-Volzhskoe parokhstva, Kronstadt, Russia; 1871 sold to Antonov, Russia, and used as tug; 1881 sold to I S Volkox, Astrakhan, Russia, and renamed **Aleksandr Volkov**; 1891 sold to D Artemen, Baku, Russia; 1892 sold to A A N Magmetov, A K K R K Ogly & A K U Mekhmed, Baku, Russia; 1892 lost in early April in Caspian Sea, on voyage from Persia to Baku.
Pioneer	1847-1851	197	Paddle steamer built by Barr & McNab, Paisley, 1844, as **Pioneer** for Bute Steam Packet Co (subsidiary of Glasgow, Paisley & Greenock Railway Co); 1847 sold to Glasgow and Liverpool Steam Shipping Co; 1851 ceded to David Hutcheson & Co, Glasgow; 1874 lengthened by 27 feet; 1879 to David MacBrayne, Glasgow; 1893 laid up; 1895 scrapped.
Pilot	1847-1851	133	Paddle steamer built by Barr & McNab, Paisley, 1844, as **Pilot** for Bute Steam Packet Co (subsidiary of Glasgow, Paisley & Greenock Railway Co); 1847 sold to Glasgow and Liverpool Steam Shipping Co; 1850 sold to William Denny & Brothers, Dumbarton; 1851 sold to Dugald Rose, William F Johnstone, John Ballardie and John Anderson, Glasgow; 1852 sold to Patrick M Halley, William F Johnstone, John Ballardie & John Anderson, Glasgow; 1853 sold to John Anderson, William F Johnstone and John Ballardie, Glasgow; 1853 sold to William F Johnstone & John Ballardie, Glasgow; 1853 sold to Robert Henderson, Robert Workman, James Denny, Dumbarton, & others, Belfast; 1862 sold to T Keenan; 1863 scrapped.

Petrel	1847-1851	192	Paddle steamer built by Barr & McNab, Paisley 1845, as **Petrel** for Bute Steam Packet Co (subsidiary of Glasgow, Paisley & Greenock Railway Co); 1847 sold to Glasgow and Liverpool Steam Shipping Co; 1851 sold to William Denny and Brothers, Dumbarton; 1851 sold to John Anderson and Patrick M Halley, Glasgow; 1853 sold to Patrick M Halley, Glasgow; 1853 sold to Patrick M Halley and William F Johnstone, Glasgow; 1855 William F Johnstone's assets sequestrated. Sold to Patrick M Halley, Glasgow; 1858 sold to Alexander Watson, Glasgow; 1860 sold to Alexander Watson and Henry Sharp, Glasgow; 1864 sold to Alexander Watson; 1865 sold to Alexander Watson and Henry Sharp, Glasgow; 1866 sold to John Dore and John Brand, following the death of Alexander Watson; 1870 sold to John Brand and Henry Sharp; c.1875 scrapped.
Maid of Perth	1847-1851		Horse drawn track boat built for service on Monklands Canal, possibly as **Thornwood**; 1847 sold to Glasgow and Liverpool Steam Shipping Co; 1851 ceded to David Hutcheson & Co; 1866 withdrawn from service, following the building of ss **Linnett**, but left in Crinan Canal for many years afterwards.
Sunbeam	1847-1851		Horse drawn track boat built at Blackwood, 1847: sold to Glasgow and Liverpool Steam Shipping Co; 1851 ceded to David Hutcheson & Co, Glasgow; 1866 withdrawn from service but left in Crinan Canal.
Cygnet	1848-1851	107	Paddle steamer built by John Reid & Co, Port Glasgow, 1848, as **Ben Nevis** for Glasgow and Liverpool Steam Shipping Co; launched as **Cygnet**; 1851 ceded to David Hutcheson & Co, Glasgow; 1879 to David MacBrayne, Glasgow; 18 September 1882 wrecked in Loch Ailort, on voyage from Glasgow to the West Highlands.
Plover	1848-1851	141	Paddle steamer built by Thomas Wingate & Co, Whiteinch, Glasgow, 1848, as **Plover** for Glasgow & Liverpool Steam Shipping Co; 1851 boiler exploded while berthed at the Broomielaw, and sold to William Denny & Brothers, Dumbarton; 1852 sold to William Whelan (as nominee for Lancashire & Furness Railway Co/ Preston & Wyre Railway Co), Lancaster; 1859 sold to Michael R Ryan, Limerick; 1862 scrapped.
Lapwing	1848-1851	110	Paddle steamer built by John Reid & Co, Port Glasgow, 1848, as **Lapwing** for Glasgow and Liverpool Steam Shipping Co; 1851 ceded to David Hutcheson & Co, Glasgow; 22 February 1859 sank after collision in fog with **Islesman**, off Kintyre between Sanda and Campbeltown, on voyage from Glasgow to Inverness.
Curlew	1849-1851	77	Paddle steamer built by David Napier, Glasgow, 1837, as **Loch Lomond** for David Napier for service on Loch Lomond; 1846 sold to H & William Ainslie, Glasgow, and renamed **Glencoe** for service on Caledonian Canal; 1849 renamed **Curlew**; 1849 sold to Glasgow and Liverpool Steam Shipping Co; 1851 ceded to David Hutcheson & Co, Glasgow; 1854 sold to Ellis Hughes, Cornelius Lancaster and George Lloyd, Liverpool, and used on Tranmere ferry; 1854 sold to Ellis Hughes and Cornelius Lancaster, Liverpool; 1855 sold to Ellis Hughes, Cornelius Lancaster and William Hey, Liverpool; 1865 scrapped.
Queen of Beauty	1849	140	Paddle steamer built by Thomas Wingate & Co, Glasgow, 1844, as **Queen of Beauty** for James Smith, James Gourlay, John Kibble, Robert Knox, James Reid, William Richmond, Robert Fleming and James Dunn, Glasgow; 1844 sold to James Gourlay, John Kibble, Robert Knox, James Reid, William Richmond and James Dunn, Glasgow; 1845 sold to William Ainslie, Fort William; 1849 sold to Glasgow and Liverpool Steam Shipping Co; 1849 register closed; 1850 sold to William Denny & Brothers, Dumbarton, and rebuilt as **Merlin**; 1851 sold to William Denny & Brothers, Dumbarton, and others; 1852 sold to William F Johnston; 1856 sold to John O Lever, Manchester; 1871 sold abroad, no further details.

Name			Comments
Maid of Morven	1849-1850	53 net	Paddle steamer built by John Wood & James Barclay, Port Glasgow, 1826, as ***Maid of Morven*** for Maid of Morven Steam Boat Co, Glasgow; 1835 sold to Robert Napier, Glasgow; 1841 sold to Alexander McKenzie, Jr, Kessock Ferry; 1846 sold to William Ainslie, Fort William; 1849 sold to Glasgow and Liverpool Steam Shipping Co; 1850 scrapped.
Lochfine	1850-1851	83	Built by William Denny & Brothers, Dumbarton, 1847, as ***Lochfine*** for William Roxburgh (Glasgow and Lochfyne Steam Packet Co), Glasgow; 1850 sold to Glasgow and Liverpool Steam Shipping Co; 1851 sold to William Denny & Brothers, Dumbarton; 1851 sold to Dumbarton Steam Boat Companies; 1862 sold to R Lang; 1864 sold to Archibald Denny and others, Glasgow, and renamed ***Loch Fyne***; 1866 sold to John McMillan & others, Glasgow; 1875 sold to John Price and Donald McFarlane, Dumbarton; 21 November 1888 collided with Jamaica Street Bridge and sank, later refloated; 1895 to Dumbarton Steamboat Co. Ltd; 1896 scrapped.

G & J Burns (1842 - 1904) – Liverpool and Irish services

Name	G & J Burns Service	Gross tons	Comments
Thetis	1845-1851	345	Paddle steamer built by Robert Napier & Co, Govan, 1845, as ***Thetis*** for Glasgow and Liverpool Steam Shipping Co; 1851 sold to William Denny & Brothers, Dumbarton; 1853 sold overseas; 1858 sold to Rigaer Dampfschiffahrt Gesellschaft, Riga; Estonia; 28 Jun 1859 grounded on Abro island, off Arensburg (now Abruka, off Kuressaare, Estonia) on voyage St Petersburg to Riga.
Orion	1847-1850	899	Paddle steamer built by Caird & Co, Greenock, 1847, as ***Orion*** for Glasgow and Liverpool Steam Shipping Co; 18 June 1850 wrecked on Barnaugh Rock, Rhinns of Galloway, off Portpatrick, on voyage from Liverpool to Glasgow.
Lyra	1848-1853	494	Paddle steamer built by Robert Napier, Govan, 1848, as ***Lyra*** for Glasgow and Liverpool Steam Shipping Co; 1849 lengthened by 19 feet; 1853 sold to William Denny & Bros, Dumbarton; 1853 sold to Patrick Gilmour, Londonderry; 1857 to Christine Gilmour, Londonderry, following death of Patrick Gilmour; 1857 sold to Londonderry Steam Boat Co Ltd, Londonderry; 13 February 1861 wrecked on Oyster Bank, Shell Rock, near Fleetwood, on voyage from Belfast to Morecambe.
Camilla	1849-1853	529	Paddle steamer built by Caird & Co, Greenock, 1849, as ***Camilla*** for H Patten & A Burrell, Greenock; 1849 bought at auction by Glasgow and Liverpool Steam Shipping Co; 1853 sold to Royal Mail Steam Packet Co, London; 1859 sold to Government of Argentina, no further details.
Laurel	1850-1854	429	Paddle steamer built by Caird & Co, Greenock, 1850, as ***Laurel*** for Glasgow & Londonderry Steam Packet Co, and completed as ***Laurel*** for Glasgow and Liverpool Steam Shipping Co.; 1854 sold to Humphrey John Hare & Co, Lancaster, as nominees for London & North Western Railway Co; 1867 scrapped.
Stork	1851-1857	432	Paddle steamer built by William Denny & Brothers, Dumbarton, 1851, as ***Stork*** for Glasgow and Liverpool Steam Shipping Co; 1856 sold to G & J Burns; 1857 sold to William Denny & Brothers, Dumbarton, and registered in names of James Denny & Peter Denny; 1858 sold to David Hutchinson & Co. Glasgow; 1860 sold to Italian buyer; 1861 sold to Italian Government, no further details.

Elk	1853-1859	548	Paddle steamer built by William Denny & Brothers, Dumbarton, 1853, as *Elk* for Glasgow and Liverpool Steam Shipping Co Ltd; 1856 sold to G & J Burns, Glasgow; 7 June 1859 wrecked at Groomsport, Belfast Lough, near Bangor, Co. Down, in fog, on voyage from Glasgow to Belfast.
Stag	1854-1864	548	Paddle steamer built by William Denny & Brothers, Dumbarton, 1853, as *Stag* for G & J Burns; 1864 sold to James T Caird, Greenock; 1864 sold to James T Nutt and resold as a blockade runner, renamed *Kate Gregg*; 1865 sold to Joseph Roberts, Nassau, and renamed *Stag*; 1867 sold to John N Beach, Liverpool; 1868 sold to owners in Buenos Aires, no further details.
Lynx	1854-1869	548	Paddle steamer built by William Denny & Brothers, Dumbarton, 1854, as *Lynx* for G & J Burns, Glasgow; 1865 sold to James T Caird, Greenock; 1869 sold to Christopher R M Talbot, MP, and converted to private yacht; 1890 to Emily J Talbot, following the death of Christopher R M Talbot; 1892 sold to Oliver S S Piner and Benjamin B David, Port Talbot; 1892 sold to John S Turnbull and scrapped.
Plover	1854	529	Built by William Denny & Brothers, Dumbarton, 1854, for Cunard Steamship Co, but completed as *Plover* for G & J Burns; 1854 sold to West Hartlepool Steam Navigation Co and registered in names of Ralph W Jackson and Robinson Watson, Stockton on Tees, and renamed *Ward Jackson*; 1862 to William S Leng & Alexander K Curtiss, West Hartlepool; 1863 sold to J C F Lencke, Germany, and renamed *Herman*; 1863 sold to William G Bennett, West Hartlepool, and renamed *May Queen*; 1863 sold to West Hartlepool Steam Navigation Co and registered in names of William S Leng and Alexander K Curtiss, West Hartlepool; 1863 sold to Joseph Spence, London; 1864 sold to John Pile; 1864 sold to Joseph Spence; 1864 sold to John Pile and Joseph Spence (Pile, Spence & Co); 1865 to Pile, Spence & Co Ltd; 1866 sold to John A Woods and Christopher D Barker, Newcastle-on-Tyne; 1867 sold to James Park, London; 1868 sold to Thomas H Pile, London; 9 April 1874 sailed from River Tyne on voyage to Cartagena, presumed foundered.
Beaver	1854-1856	365	Built by Robert Steele & Co, Greenock, 1854, as *Beaver* for G & J Burns; 1856 sold to Burns & MacIver, Glasgow, and re-sold to Peter Denny, jnr, Dumbarton; 1857 sold to Meinhard E Robinow; 1859 sold to Malcomson Brothers, Waterford; 1861 sold overseas, no further details.
Zebra	1855-1856	791	Built by Caird & Co, Greenock, 1855, as *Zebra* for G & J Burns; 22 July 1856 aground at Parnvoose Cove on the Lizard, Cornwall, on voyage from Le Havre to Liverpool and wrecked.
Otter	1855	473	Built by Robert Steele & Co, Greenock, 1855, as *Plover* and completed for G & J Burns as *Otter*; 1855 sold to Robert Steele, Robert Steele, jnr, and William Steele (Robert Steele & Co), Greenock, and immediately resold to John Ormston & John T Dobson, Newcastle; 1864 to Tyne Steam Shipping Co Ltd, Newcastle; 20 February 1873 foundered in the Wold, off Yarmouth, in thick fog, on voyage from Newcastle to Antwerp.
Panther	1857-1859	701	Paddle steamer built by Robert Steel & Co, Greenock, 1857, as *Panther* for G & J Burns; 1859 sold to Peter Denny and John McAusland, Dumbarton; 1860 sold to the Italian Government at Palermo; 1861 to Royal Italian Navy and renamed *Plebiscito* as transport; 1875 disposed of.
Leopard	1858-1860	691	Paddle steamer built by William Denny & Brothers, Dumbarton, 1858, as *Leopard* for G & J Burns; 1860 sold to Peter Denny and John McAusland, Dumbarton; 1862 sold to James A K Wilson, Liverpool, (Fraser, Trenholm & Co), re-sold to George D Harris, Nassau, Bahamas; 1863 sold for use as a blockade runner and renamed *Stonewall Jackson*; 12 April 1863 ran aground off Charleston and set on fire to avoid capture, on voyage to Charleston.

Name	Years	No.	Details
Harrier	1858-1860	384	Built by Alexander Denny, Dumbarton, 1858, as **Harrier** for G & J Burns; 1860 sold to J & G Thomson and resold to Thomas S Begbie, London; 1860 sold overseas, no further details.
Heron	1860-1863	624	Built by William Denny & Brothers, Dumbarton, 1860, as **Heron** for G & J Burns; 1863 sold to George Thomson; 1864 sold to General Steam Navigation Co, London; 1889 scrapped.
Ostrich	1860-1866	624	Built by William Denny & Brothers, Dumbarton, 1860, as **Ostrich** for G & J Burns; 1866 sold to Martin Pratt and resold to General Steam Navigation Co, London; 1889 scrapped.
Giraffe	1860-1862	677	Paddle steamer built by J & G Thomson, Govan, 1860, as **Giraffe** for G & J Burns; 1862 sold to John Williamson, Liverpool, and re-sold to Edward Pembroke, London, re-sold to Confederate Government as blockade runner and renamed **Robert E Lee**; 9 November 1863 captured by USS **James Adger** off Wilmington, on voyage from Halifax, Nova Scotia; 1864 bought by US Navy from Boston Prize Court and renamed USS **Fort Donelson**; 1865 sold to George S Brown, Baltimore, and renamed **Isabella**; 1866 sold to Chilean Navy and renamed **Concepcion**; no further details.
Wolf	1863-1871	670	Paddle steamer built by Robert Napier & Sons, Govan, 1863, as **Wolf** for G & J Burns; 1871 sold to London & South Western Railway, Southampton; 1873 engines compounded; 1897 scrapped.
Roe	1863	610	Paddle steamer built by Caird & Co, Greenock, 1863, as **Roe** for J & G Burns but launched as **City of Petersburg** for George Campbell, Dunoon, and others as blockade runner; 1865 sold to George Campbell, Liverpool, re-sold to Liverpool & Dublin Steam Navigation Co Ltd, Liverpool, and renamed **Bridgewater**; 1868 sold to Alexander A Laird, Glasgow, (see page 113), and renamed **Garland**, re-sold to Glasgow & Londonderry Steam Packet Co; 1873 re-engined; 1874 sold to Edward M de Bussche, Ryde, Isle of Wight; 1876 sold under mortgage to Thomas Redway, Exmouth; 1880 sold under mortgage to Warrenpoint Steam Packet Co, Warrenpoint; 1881 sold to Donald McCall, Greenwich; 1881 sold to James A Bayley, London; 1881 scrapped.
Fox	1863	607	Paddle steamer built by Caird & Co, Greenock, 1863, as **Fox** for J & G Burns but completed for David McNutt as blockade runner **Nola**; 30 December 1863 aground on rocks north-west of Bermuda, on voyage from Clyde to Nassau and became a wreck.
Penguin	1864-1878	680	Built by Tod & McGregor, Glasgow, 1864, as **Penguin** for G & J Burns; 1878 sold to James R Thompson; 1879 sold to John Darling and resold to Union Steamship Co of New Zealand Ltd, Dunedin (intended name change to **Tarawera** never made); 1882 engines compounded; 12 February 1909 wrecked off Cape Terawhiti, during gale, on voyage from Picton to Wellington.
Roe	1864-1865	559	Paddle steamer built by Caird & Co, Greenock, 1864, as **Roe** for G & J Burns; 1865 sold to William E Hutchinson, Leicester, and William P Price, Gloucester (nominees of Midland Railway Co), Lancaster; 1868 sold to Barrow Steam Navigation Co, Barrow, and registered in names of nominees; 1887 scrapped.
Fox	1864	560	Paddle steamer built by Caird & Co, Greenock, 1864, as **Fox** for G & J Burns, sold to James T Caird (Caird & Co), Greenock, and re-sold to Crenshaw Brothers as blockade runner **Agnes E Fry**; 27 December 1864 aground off Old Inlet, North Carolina, while leaving Charleston, and wrecked.
Beagle	1864-1865	454	Built by Tod & McGregor, Glasgow, 1864, as **Beagle** for G & J Burns; 8 November 1865 sank off Cumbrae Island, on voyage from Belfast to Glasgow, following collision with Anchor liner **Napoli**.

Name	Years	Tons	Details
Buffalo	1865-1881	675	Paddle steamer built by Caird & Co, Greenock, 1865, as **Buffalo** for G & J Burns; 1881 sold to Barclay, Curle & Co, Glasgow, in part payment for new steamships. Registered in names of Archibald Gilchrist, John Ferguson and Andrew MacLean, new engines; 1881 sold to Barrow Steam Navigation Co, Barrow, and registered in names of nominees, renamed **Donegal**; 1898 sold to British & Continental Syndicate Ltd, London, resold to Alfred J Campion, London and returned to British & Continental Syndicate Ltd, London; 1898 sold by the High Bailiff of the County Court of Sussex to James G Smith, London; 1900 scrapped.
Llama	1865-1881	675	Paddle steamer built by Caird & Co, Greenock, 1865, as **Llama** for G & J Burns; 4 March 1881 grounded near Corsewall Point in snowstorm, on voyage from Belfast to Glasgow. Refloated 3 April with damage to hull, towed to Glasgow and scrapped.
Weasel	1866-1869	489	Built by J & G Thomson, Govan, 1866, as **Weasel** for G & J Burns; 1869 sold to James R Thomson and re-sold to Thomas Henderson and John Henderson (Henderson Brothers, Anchor Line), Glasgow; 1870 sold to Gluckstad & Co, Christiania, Norway, and renamed **Scotia**, re-sold to Thomas Henderson & John Henderson (Henderson Brothers, Anchor Line), Glasgow; 1873 lengthened by 68 feet; 1878 new engines; 1882 sold to James Donald, Glasgow, and re-sold to James Donald and others (Walker, Donald & Co), Glasgow; 1887 condemned at New York; 1888 sold to Richard Holman, London and renamed **Exmouth**, re-sold sold to Exmouth Steamship Ltd, re-sold to Thomas H Nelson, Liverpool; 1891 sold to Thomas H Nelson and James Smith, Liverpool; 1893 sold by James Smith to Edmund Hulbert, London; 1895 scrapped.
Camel	1866-1881	688	Paddle steamer built by Caird & Co, Greenock, 1866, as **Camel** for G & J Burns; 1881 sold to Barclay, Curle & Co, in part payment for new steamships. Registered in names of Archibald Gilchrist, John Ferguson and Andrew MacLean; new engines and resold to Barrow Steam Navigation Co, Barrow, registered in names of nominees and renamed **Londonderry**; 1904 scrapped.
Snipe	1866-1870	638	Built by J & G Thomson, Govan, 1866, as **Snipe** for G & J Burns; 1870 sold to Reuben D Sassoon, London, and re-sold to Hajee GA Sherajee Mogul, Bombay; 16 August 1871 sold overseas, and sailed 26 August from Bombay to Bushire, Persian Gulf, and believed lost.
Racoon	1868-1880	831	Paddle steamer built by J & G Thomson, Govan, 1868, as **Racoon** for G & J Burns; 1880 sold to Barclay, Curle & Co, in part payment for new steamships. Registered in names of John Thomson, Archibald Gilchrist and Andrew MacLean and resold to the Mayor and Commonalty of the City of London as cattle ship; 1907 scrapped.
Raven	1868-1881	778	Built by J & G Thomson, Govan, 1868, as **Raven** for G & J Burns; 1877 new engines; 1881 sold to Basilio Papayanni, Liverpool; 1881 sold to Hellenic Steamship Co, Syra, Greece, and renamed **Pinios**; 1894 to New Hellenic Steam Navigation Society, Syra; Greece; 1906 sold to Soc de Nav à Vapeur des Forges et Chantiers 'Neorion'; 1908 sold to Cyclades Steam Navigation Co, Greece; 1916 sold to Hellenic Co of Maritime Enterprises (A Pallios); 1927-28 sold to Mandafouni Steamship Co, Piraeus, Greece, and renamed **Themistocles**; 1929 sold to K Togias Steamship Co (E K Togias), Syra, Greece, and renamed **Karystos Togia**; 1931 sold to Hellenic Coast Lines Co Ltd, Piraeus, Greece; 1934 scrapped.
Bear	1870-1895	632	Built by J & G Thomson, Govan, 1870, as **Bear** for G & J Burns; 1895 sold to Liverpool & Clyde Steam Navigation Co Ltd (Charles MacIver & Co); 1903 sold under mortgage to Bell's Asia Minor Steamship Co Ltd, Liverpool, (E Minotto); 1906 sold to William M Moss, Liverpool and Sir John R Ellerman, London; 1907 sold to George Milne, Glasgow (shipbroker) and re-sold to Hellenic Steam Navigation Co (John MacDowall & Barbour), Piraeus, Greece, renamed **Calypso**; 1913 to Navigation Hellénique John MacDowall, Piraeus, Greece; 31 January 1914 wrecked near Paros, on voyage from Piraeus to Santorini.

Name	Years	No.	History
Bison	1871-1872	1015	Built by J & G Thomson, Govan, 1871, as **Bison** for G & J Burns; 1872 sold to Royal Mail Steam Packet Co, London, and renamed **Belize**; 1888 sold to Government of Haiti and renamed **Defense**; 1890 out of register.
Ferret	1871-1873	344	Built by J & G Thomson, Govan, 1871, as **Ferret** for G & J Burns; 1873 sold to Dingwall & Skye Railway Co, Inverness; 1878 Dingwall & Skye Railway Co acquired by Highland Railway Co; 1878 to Highland Railway Co; 1880 stolen: sailed to Australia under false names of **Benton** and **India**, recognised & seized in Australia. Resumed name of **Ferret**; 1881 sold to William Whinham (South Eastern Coast and Inter-Colonial Steam Ship Co Ltd); 1883 South Eastern Coast and Inter-Colonial Steam Ship Co Ltd acquired by Adelaide Steamship Co Ltd, Adelaide; 14 November 1920 wrecked in fog on Yorke Peninsula, on voyage from Port Adelaide to Spencer Gulf ports.
Owl	1873-1895	915	Built by Tod & McGregor, Partick, 1873, as **Owl** for G & J Burns; 1895 sold to Liverpool & Clyde Steam Navigation Co Ltd (Charles MacIver & Co); 1901 sold to Alexander Wylie, Johnstone, Renfrewshire; 1901 sold to MacDowall & Barbour, Piraeus, Greece, and renamed **Antigone**; 1907 to Hellenic Steam Navigation Co (MacDowall & Barbour), Piraeus, Greece; 1913 to Navigation Hellénique John MacDowall, Piraeus; Greece; 1917 sold to Hellenic Co of Maritime Enterprises (A Palios), Piraeus, Greece; 1918-19 name re-styled **Antigoni**; 1928 scrapped.
Hornet	1874-1877	548	Built by Blackwood & Gordon, Port Glasgow, 1874, as **Hornet** for G & J Burns; 1877 sold to James R Thomson and resold to Charles Henderson; 1878 sold to Charles Henderson and William Wilson; 1881 sold to William Pearce, Govan, and lengthened by 32 feet; 27 January 1884 sank in force 9 gale off Lundy Island, on voyage from Newport, Monmouthshire, to Marseilles.
Wasp	1874-1888	550	Built by Blackwood & Gordon, Port Glasgow, 1874, as **Wasp** for G & J Burns; 1878 sold to M. Langlands & Sons and G & J Burns, and re-sold to M Langlands & Sons; 1879 sold to M Langlands & Sons, G & J Burns & Robert Gilchrist & Co, managed by M Langlands & Sons; 11 July 1888 sank in River Mersey after a collision with the anchored Norwegian barque **Hypatia**, on voyage from Glasgow to Liverpool. Wreck dispersed with explosives.
Jackal	1875	438	Built by A & J Inglis, Pointhouse, 1875, as **Jackal** for G & J Burns, but completed as **Taiaroa** for Albion Shipping Co, Glasgow, for New Zealand coastal service. Registered in name of James Galbraith; 1878 sold to Union Steamship Co of New Zealand Ltd, Dunedin; 11 April 1886 wrecked at the entrance to the Clarence River, Waipapa Point, Kaikoura Coast, New Zealand, on voyage from Wellington to Lyttelton.
Walrus	1878-1880	870	Built by J & G Thomson, Dalmuir, 1878, as **Walrus** for G & J Burns; 1880 sold to Royal Hellenic Navy and renamed **Kanaris**; 1889 renamed **Psara**; 1926 sold to Geo Domestinis & Co, Piraeus, Greece; 1930 sold to Hellenic Coast Lines Co Ltd, Piraeus, Greece; 1933 scrapped.
Mastiff	1878-1906	871	Built by J & G Thomson, Clydebank, 1878, as **Mastiff** for G & J Burns; 1905 to G & J Burns Ltd; 1906 sold to Joseph Guggero, Gibraltar, and re-sold to M H Bland & Co Ltd, Gibraltar, renamed **Gibel Dersa**; 1924 scrapped.
Rook	1878-1881	430	Built by J & G Thomson, Clydebank, 1878, as **Moorfowl** for G & J Burns, completed as **Rook**; 1881 sold to Napier & Co, Glasgow; 1881 sold to Napier Shipping Co Ltd, Glasgow; 1883 sold to Charles Barrie, Dundee and resold to G W Nicoll & Co, Sydney, Australia; 1884 sold to Government of South Australia and renamed **Palmerston**; 1886 sold to J Darling, jnr, Port Adelaide, Australia; 1888 sold to Mount Pleasant Coal & Iron Mining Co Ltd; Sydney, Australia; 1928 sold to H P Stacey, Sydney, Australia; 1929 sold to Roger J Butler, Sydney, Australia, and converted to trawler; 29 May 1929 sank following collision with trawler **Millimumul**, off Jervis Bay.

Alligator	1881-1907	932	Built by Barclay, Curle & Co Ltd, Whiteinch, 1881, as **Alligator** for G & J Burns; 1905 to G & J Burns Ltd; 1907 sold to Alexander W Wylie, Johnstone, and resold to Hellenic Steam Navigation Co (MacDowall & Barbour), Piraeus, Greece, and renamed **Ismene**; 1913 to Navigation Hellénique John MacDowall, Piraeus, Greece; 1917 sold to Hellenic Co of Maritime Enterprises (A Palios), Piraeus, Greece; 1927-28 sold to Alexandros G Yannoulatos, Piraeus, Greece, and renamed **Afovos**; 1929 to Hellenic Coast Lines Co Ltd, Piraeus, Greece; 1933 renamed **Rodos**; 1934 scrapped.
Dromedary	1881-1909	922	Built by Barclay, Curle & Co Ltd, Whiteinch, 1881, as **Dromedary** for G & J Burns; 1905 to G & J Burns Ltd; 1909 sold to Alexander D Brown, St. John's, Newfoundland, and re-sold to the Reid Newfoundland Co Ltd, St. John's, Newfoundland, renamed **Invermore**; 10 July 1914 wrecked at Brig Harbour, Point, Labrador, on voyage from Saint John's to Labrador.
Gorilla	1881-1913	922	Built by Barclay, Curle & Co Ltd, Whiteinch, 1881, as **Gorilla** for G & J Burns; 1905 to G & J Burns Ltd; 1913 sold to Navigation à Vapeur 'Ionienne' G Yannoulatou Frères, Piraeus, Greece, and renamed **Nafkratousa**; 30 January 1918 caught fire at Gallipoli, Italy. Towed out into North Bay and sunk by gunfire from a British naval motor launch.
Lizard	1881-1883	411	Built by Blackwood & Gordon, Port Glasgow, 1881, as **Lizard** for G & J Burns; 1883 sold to Navigation à Vapeur Egée (P M Courtgi & Co), Constantinople, Turkey, and renamed **Crete**; 1904 sold to Destouni & Yannoulatos, Piraeus, Greece, and renamed **Kerkyra**; 1912 sold to Navigation à Vapeur 'Ionienne' G Yannoulatou Frères, Piraeus, Greece; 19 June 1917 torpedoed by submarine U4 and sank in the Gulf of Taranto, 12 miles SW of Gallipoli, on voyage from Taranto to Corfu.
Locust	1881-1883	411	Built by Blackwood & Gordon, Port Glasgow, 1881, as **Locust** 1881 for G & J Burns; 1883 sold to Navigation à Vapeur Egée (P M Courtgi & Co, Constantinople, Turkey, and renamed **Chios**; 1904 sold to Destouni & Yannoulatos, Piraeus, Greece, and renamed **Zakynthos**; 1912 sold to Navigation à Vapeur 'Ionienne' G Yannoulatou Frères, Piraeus, Greece; 1924-25 to S A Ionienne de Navigation à Vapeur Yannoulatos, Piraeus, Greece; 29 September 1929 wrecked at Zemala Bay, Ayios Pavlos, Lemnos.
Lamprey	1881-1898	311	Built by Blackwood & Gordon, Port Glasgow, 1881, as **Lamprey** for G & J Burns; 1898 sold to Benjamin Rouxel, Le Légué, France, and renamed **Hirondelle**; 3 September 1906 sank near Perros, France, after touching rocks at Sept Iles, in fog, on voyage from Le Havre to Brest.
Seal	1882-1907	678	Built by A & J Inglis, Pointhouse, 1877, as **North Eastern** for Ardrossan Shipping Co, Ardrossan; 1882 sold to G & J Burns and renamed **Seal**; 1905 to G & J Burns Ltd; 1907 sold to Unione Austriatica di Nav. S A (Fratelli Cosulich), Trieste, Austria, and renamed **Eleni**; 1908 sold to Achaia Steamship Co Ltd (W Morphy & Sons and Crowe & Stephens), Patras, Greece; 31 August 1917 captured by submarine and scuttled with explosives in the Kassos Strait, on voyage from Volos to Alexandria.
Grampus	1882-1907	686	Built by A & J Inglis, Pointhouse, 1877, as **North Western** for Ardrossan Shipping Co, Ardrossan; 1882 sold to G & J Burns and renamed **Grampus**; 1905 to G & J Burns Ltd; 1907 sold to Unione Austriatica di Nav. S A (Fratelli Cosulich), Trieste, Austria, and renamed **Elda**; 1908 sold to Achaia Steamship Co Ltd (W Morphy & Sons and Crowe & Stephens), Patras, Greece; 1919-20 sold to National Steam Navigation Co Ltd of Greece Embiricos Bros), Andros, Greece and renamed **Tynos**; 1924-25 sold to Steam Navigation of Samos (D Inglessi Fils), Samos, Greece, and renamed **Tinos**; 1928-29 to D Inglessi Fils S A de Nav. de Samos, Samos, Greece; 1936 scrapped.

Limpet	1882-1889	475	Built by Barclay, Curle & Co Ltd, Whiteinch, 1882, as *Limpet* for G & J Burns; 1889 sold to Sligo Steam Navigation Co, Sligo; 1890 renamed *Sligo*; 5 February 1912 wrecked at Ardbowline Island, Donegal Bay, on a voyage from Garston to Sligo.
Buzzard	1884-1887	831	Built by J & G Thomson, Clydebank, 1884, as *Buzzard* for G & J Burns; 1887 sold to Hellenic Steamship Co, Syra, Greece, and renamed *Sfactiria*; 1894-95 to New Hellenic Steamship Soc, Syra, Greece; 1900 sold to Royal Hellenic Navy; 1921 disposed of.
Hare	1887-1889	771	Built by Barclay, Curle & Co Ltd, Whiteinch, 1886, as *Hare* for G & J Burns; 1889 sold to John P Tocque, St. Helier, Jersey, and re-sold to David J Stewart, Dublin, & George Lowen, Manchester (Dublin & Manchester Steamship Co), Dublin; 1910 sold to George Lowen, Manchester; 14 December 1917 torpedoed by submarine U62 and sank 7 miles E of Kish light vessel, on voyage from Manchester to Dublin.
Cobra	1889	847	Paddle steamer built by Fairfield Shipbuilding & Engineering Co Ltd, Govan, 1889, as *Cobra* for G & J Burns. Registered under Richard Barnwell, Fairfield Works, as nominee for Fairfield Shipbuilding & Engineering Co Ltd; 1890 renamed *St Tudno* and operated by New North Wales Steamship Co; 1890 renamed *Cobra*; 1891 sold to Nordsee Dampfschiff Gesellschaft, Hamburg, Germany; 1905 sold to Hamburg-Amerika Linie, Hamburg; 1918 acquired by French Government; 1922 sold to Mahr & Beyer and scrapped.
Adder	1890-1906	976	Paddle steamer built by Fairfield Shipbuilding & Engineering Co Ltd, Govan, 1890, as *Adder* for G & J Burns; 1905 to G & J Burns Ltd; 1906 sold to Alfred S Chovil, Birmingham, and resold to S Lambruschini, Buenos Aires, Argentina, and renamed *Rio de la Plata*; 1914 to Cia de Navegacion S Lambruschini Ltda, Buenos Aires, Argentina; 24 December 1918 wrecked at Maldonado, on voyage from Buenos Aires to Santos.
Grouse	1891-1922	386	Built by Caird & Co Ltd, Greenock, 1891, as *Grouse* for G & J Burns; 1905 to G & J Burns Ltd; 1920 G & J Burns Ltd acquired by Coast Lines Ltd; 1922 owner renamed Burns & Laird Lines Ltd; 1922 sold to Grahamston Shipping Co Ltd (T L Duff & Co), and renamed *Kelvindale*; 1923 Grahamston Shipping Co Ltd acquired by Burns & Laird Lines Ltd; 1924 to Coast Lines Ltd, Liverpool, and renamed *Denbigh Coast*; 1929 sold to David MacBrayne (1928) Ltd, Glasgow. and renamed *Lochdunvegan*; 1934 owner renamed David MacBrayne Ltd; 1948 scrapped.
Hound	1893-1922	1,061	Built by Fairfield Shipbuilding & Engineering Co Ltd, Govan, 1893, as *Hound* for G & J Burns; 1905 to G & J Burns Ltd.; 1920 G & J Burns Ltd. acquired by Coast Lines Ltd; 1922 owner renamed Burns & Laird Lines Ltd; 1925 sold to M G A Manuelides Brothers, Piraeus, Greece, and renamed *Mary M*; 1926 to Manuelides Nav Co Ltd, Piraeus, Greece; 1929 Manuelides Nav Co Ltd acquired by Hellenic Coast Lines Co Ltd; 1929 to Hellenic Coast Lines Co Ltd, Piraeus, Greece; 1933 renamed *Lesbos*; 1942 renamed *Korytza*; 1944 acquired by Ministry of War Transport, London; 1946 returned to Greek owner, Piraeus; Greece; 1950 scrapped.
Spaniel	1895-1921	1,116	Built by A & J Inglis Ltd, Pointhouse, 1895, as *Spaniel* for G & J Burns; 1905 to G & J Burns Ltd; 1920 G & J Burns Ltd acquired by Coast Lines Ltd.; 1921 to Belfast Steamship Co Ltd, Belfast, and renamed *Caloric*; 1932 scrapped.
Pointer	1896-1922	1,130	Built by A & J Inglis Ltd, Pointhouse, 1896, as *Pointer* for G & J Burns; 1905: to G & J Burns Ltd; 1920 G & J Burns Ltd acquired by Coast Lines Ltd; 1922 owner renamed Burns & Laird Lines Ltd; 1929 renamed *Lairdsvale*; 1933 scrapped.

Name	Service	Gross tons	Comments
Ape	1898-1912	467	Built by Barclay, Curle & Co Ltd, Whiteinch, 1898, as *Ape* for G & J Burns; 1905 to G & J Burns Ltd; 1912 sold to H Newhouse & Co Ltd, Norwich (Henry Newhouse), Yarmouth; 1915 to Newhouse Ltd, Norwich (Henry Newhouse), Yarmouth; 1916 sold to Henry Newhouse, Norwich, Yarmouth, and re-sold to Cunningham, Shaw & Co Ltd, London (Vernon S Lovell); 1918 sold to William Y McDonald, George Davidson and William Merrylees, Aberdeen; 1918 sold to North of Scotland & Orkney & Shetland Steam Navigation Co Ltd, Aberdeen, and renamed *Fetlar*; 13 April 1919 wrecked on Bunel Rocks, near St Malo, on voyage from Southampton to St Malo.
Boar/Magpie	1898-1922	1,227	Built by A & J Inglis Ltd, Pointhouse, 1898, as *Boar* but completed as *Magpie* for G & J Burns; 1905 to G & J Burns Ltd; 1920 G & J Burns Ltd acquired by Coast Lines Ltd; 1922 owner renamed Burns & Laird Lines Ltd; 1929 renamed *Lairdsgrove*; 1948 sold to Metal Industries Ltd, Faslane, as an accommodation ship; 1950 scrapped.
Vulture	1898-1922	1,168	Built by A & J Inglis Ltd, Pointhouse, 1898, as *Vulture* for G & J Burns; 1905 to G & J Burns Ltd; 1920 G & J Burns Ltd acquired by Coast Lines Ltd; 1922 owner renamed Burns & Laird Lines Ltd; 1929 renamed *Lairdsrock*; 1937 sold to David MacBrayne Ltd, Glasgow, and renamed *Lochgarry*; 21 January 1942 foundered off Rathlin Island, on voyage from Glasgow to Oban.

G & J Burns Limited (1904 - 1922)

Name	G&J Burns Ltd Service	Gross tons	Comments
Viper	1906-1920	1,713	Built by Fairfield Shipbuilding & Engineering Co Ltd, Govan, 1906, as *Viper* for G & J Burns Ltd; 1920 G & J Burns Ltd acquired by Coast Lines Ltd; 1920 sold to Isle of Man Steam Packet Co Ltd, Douglas, and renamed *Snaefell*; 1945 scrapped (not completed until 1948).
Woodcock	1906-1922	1,523	Built by John Brown & Co Ltd, Clydebank, 1906, as *Woodcock* for G & J Burns Ltd; 1914 requisitioned as HMS *Woodnut*; 1919 de-commissioned and refitted; 1920 returned to G & J Burns Ltd as *Woodcock*; G & J Burns Ltd acquired by Coast Lines Ltd; 1922 owner renamed Burns & Laird Lines Ltd; 1929 renamed *Lairdswood*; 1930 sold to the Aberdeen Steam Navigation Co Ltd, Aberdeen; 1931 renamed *Lochnagar*; 1946 sold to Rena Cia de Naviera S. A. (Panos Protopapos, Alexandria, Egypt, and renamed *Rena*; 1952 renamed *Blue Star* and scrapped.
Partridge	1906-1922	1,523	Built by John Brown & Co Ltd, Clydebank, 1906, as *Partridge* for G & J Burns Ltd; 1915 requisitioned as HMS *Partridge II*; 1920 returned to G & J Burns Ltd as *Partridge*; 1920 G & J Burns Ltd acquired by Coast Lines Ltd; 1922 owner renamed Burns & Laird Lines Ltd; 1929 renamed *Lairdsloch*; 1936 scrapped.
Lurcher	1906-1920	993	Built by Scott's Shipbuilding & Engineering Co Ltd, Greenock, 1906, as *Lurcher* for G & J Burns Ltd; 1920 G & J Burns Ltd. acquired by Coast Lines Ltd; 1920 to British & Irish Steam Packet Co Ltd, Dublin, and renamed *Lady Meath*; 1925 to City of Cork Steam Packet Co Ltd, Cork, and renamed *Inniscarra*; 1935 sold to Wexford Steamships Co Ltd, Wexford, and renamed *Menapia*; 1939 scrapped.
Setter	1906-1918	993	Built by Scott's Shipbuilding & Engineering Co Ltd, Greenock, 1906, as *Setter* for G & J Burns Ltd; 13 September 1918 torpedoed by submarine U64 and sank, 6 miles NW by N. of Corsewall Point, on voyage from Manchester to Glasgow.
Redbreast	1908-1917	1,313	Built by A & J Inglis Ltd, Pointhouse, 1908, as *Redbreast* for G & J Burns Ltd.; 1915 requisitioned, later to become fleet messenger ship; 15 July 1917 torpedoed by submarine U38 and sank in Aegean Sea.

Name	Service	Gross tons	Comments
Sable	1911-1922	687	Built by A & J Inglis Ltd, Pointhouse, 1911, as *Sable* for G & J Burns Ltd.; 1920 G & J Burns Ltd acquired by Coast Lines Ltd; 1922 owner renamed Burns & Laird Lines Ltd; 1929 renamed *Lairdselm*; 22 December 1929 sank while sheltering in Loch Ryan, during a voyage from Glasgow to Belfast.
Ermine	1912-1917	1,777	Built by Fairfield Shipbuilding & Engineering Co Ltd, Govan, 1912, as *Ermine* for G & J Burns Ltd; 1915 requisitioned as fleet messenger ship; 2 August 1917 mined and sank in Aegean Sea, off Stavros, on a voyage from Stavros to Mudros. Mine laid by German submarine U23.
Coney	1918-1922	697	Built by Harland & Wolff Ltd, Govan, 1918, as *Coney* for G & J Burns Ltd; 1920 G & J Burns Ltd acquired by Coast Lines Ltd; 1922 owner renamed Burns & Laird Lines Ltd; 1929 renamed *Lairdsferry*; 1952 scrapped.
Moorfowl	1919 1920-1922	1,464	Built by A & J Inglis Ltd, Pointhouse, 1919, as *Moorfowl* for G & J Burns Ltd, acceptance refused and completed as *Killarney* for City of Cork Steam Packet Co Ltd, Cork; 1920 to G & J Burns Ltd (G & J Burns Ltd acquired by Coast Lines Ltd) and renamed *Moorfowl*; 1922 to Burns & Laird Lines Ltd; 1929 renamed *Lairdsmoor*; 7 April 1937 sunk following collision with Shaw, Savill & Albion's *Taranaki* off Mull of Galloway, on voyage from Dublin to Glasgow.
Setter	1920	1,217	Built by William Beardmore & Co Ltd, Dalmuir, 1920, as *Whippet*, but completed as *Setter* for G & J Burns Ltd, G & J Burns Ltd acquired by Coast Lines Ltd; 1920 to British & Irish Steam Packet Co Ltd, Dublin, and renamed *Lady Kildare*; 1931 to Belfast Steamship Co Ltd, Belfast; 1932 renamed *Ulster Castle*; 1950 scrapped.
Redbreast	1921-1922	772	Built by Harland & Wolff Ltd, Govan, 1920, as *Princess Caroline* for M Langlands & Sons Ltd, but completed 1921 as *Redbreast* for G & J Burns Ltd; 1922 to Burns & Laird Lines Ltd; 1926 to Coast Lines Ltd, Liverpool, and renamed *Sutherland Coast*; 1930 to Burns & Laird Lines Ltd and renamed *Lairdsbrook*; 1960 scrapped.
Gorilla	1922	772	Built by Harland & Wolff Ltd, Govan, 1922, as *Princess Dagmar* for M Langlands & Sons Ltd, but completed as *Gorilla* for G & J Burns Ltd; 16 August 1922 sunk in channel at Cork by Irish rebels. Raised and reconditioned by Harland & Wolff, Govan, to Burns & Laird Lines Ltd; 1925 to Coast Lines Ltd, Liverpool, and renamed *Cumberland Coast*; 1929 to City of Cork Steam Packet Co Ltd, Cork, and renamed *Kinsale*; 1933 to Coast Lines Ltd, Liverpool, and renamed *Cambrian Coast*; 1947 to Belfast Steamship Co Ltd, Belfast, and renamed *Ulster Merchant*; 1954 scrapped.
Lurcher	1922	774	Built by A & J Inglis Ltd, Pointhouse, 1922, for M Langlands & Sons Ltd, but completed as *Lurcher* for Burns & Laird Lines Ltd; 1925 to Coast Lines Ltd, Liverpool, and renamed *Scottish Coast*; 1938 to Belfast Steamship Co Ltd, Belfast, and renamed *Ulster Coast*; 1954 sold to Ahern Shipping Ltd, Montreal, Canada, and renamed *Ahern Trader*; 10 January 1960 wrecked off Gander Bay, Canada.

Burns Steamship Company Limited (1908-1922)

Name	Burns SS Service	Gross tons	Comments
Wren	1908-1913	817	Built by A & J Inglis Ltd, Pointhouse, Glasgow, 1885, as *General Gordon* for Dublin & Glasgow Sailing and Steam Packet Co, Dublin; 1895 renamed *Duke of Gordon*; 1908 Dublin & Glasgow Sailing and Steam Packet Co acquired by G & J Burns Ltd, to Burns Steamship Co Ltd, Glasgow, and renamed *Wren*; new engines; 1913 sold to Administration de Nav à Vap. Ottomane, Constantinople, Turkey, and renamed *Eureuk*; 1919-20 name re-styled *Yeureuk*; 18 April 1920 wrecked at Antchiros, 90 miles from Samsun, on voyage from Samsun to Istanbul.

Sparrow	1908-1910	997	Built by Ailsa Shipbuilding Co Ltd, Troon, 1892, as **Duke of Fife** for Dublin & Glasgow Sailing and Steam Packet Co, Dublin; 1908 Dublin & Glasgow Sailing and Steam Packet Co acquired by G & J Burns Ltd, to Burns Steamship Co Ltd, Glasgow, and renamed **Sparrow**; 1910 sold to Navigation Orientale (P Pantaleon), Smyrna, Greece, and renamed **Arcadia**; 1922 to Navigation Pantaleon (P Pantaleon Fils), Piraeus, Greece; 1929 to Hellenic Coast Lines Co Ltd, Piraeus, Greece; 1933 renamed **Chios**; 18 April 1941 bombed and sank in air attack at Eretria, near Chalkis, Greece.
Puma	1908-1922	1,226	Built by Caledon Shipbuilding & Engineering Co Ltd, Dundee, 1899, as **Duke of Rothesay** for Dublin & Glasgow Sailing and Steam Packet Co, Dublin; 1908 Dublin & Glasgow Sailing and Steam Packet Co acquired by G & J Burns Ltd; to Burns Steamship Co Ltd, Glasgow, and renamed **Puma**; 1920 G & J Burns Ltd and subsidiary Burns Steamship Co Ltd acquired by Coast Lines Ltd; 1922 to Burns & Laird Lines Ltd.; 1929 renamed **Lairdsford**; 1934 scrapped.
Tiger	1908-1922	1,389	Built by Caledon Shipbuilding & Engineering Co Ltd, Dundee, 1906, as **Duke of Montrose** for Dublin & Glasgow Sailing and Steam Packet Co, Dublin; 1908 Dublin & Glasgow Sailing and Steam Packet Co acquired by G & J Burns Ltd; to Burns Steamship Co Ltd, Glasgow, and renamed **Tiger**; 1920 G & J Burns Ltd. and subsidiary Burns Steamship Co Ltd acquired by Coast Lines Ltd; 1922 to Burns & Laird Lines Ltd; 1929 renamed **Lairdsforest**; 1930 to British & Irish Steam Packet Co Ltd, Dublin, converted to cattle carrier and renamed **Lady Louth**; 1934 scrapped.

INDEX

Ship name index for Laird and Burns vessels with date of build in brackets

Achilles (1839) 61, 64, 120	*Dromedary* (1881) 80, 89, 98, 129	*Hornet* (1874) 77, 78, 128
Actæon (1837) 61, 119	*Dunoon Castle* (1826) 67, 121	*Hound* (1893) 87, 94, 95, 99, 103, 130
Adder (1890) 85-87, 93-95, 130	*Dunrobin Castle* (1846) 67, 122	*Inverary Castle* (1839) 67, 121
Ailsa Craig **(1825)** 5, 6, 59, 61, 118	*Duntroon Castle* (1842) 67, 121	*Inverness* (1832) 62, 64, 120
Alligator (1881) 80, 89, 96, 98, 129	*Dunure* (1878) 51, 55, 116	*Iris* (1876) 30, 32-35, 113
Antelope (1833) 12, 60, 64, 117		*Irishman* (1854) 21, 22, 26-28, 30, 31, 111
Ape (1898) 89, 93, 131	*Eagle* (1835) 60, 119	
Arbutus (1854) 18, 21, 23, 109	*Eagle* (1857) 17, 112	*Ivy* (1888) 38, 41, 43, 44, 115
Arbutus (1875) 30-32, 113	*Eclipse* (1826) 11, 59, 60, 117	
Argyle (1815) 10, 11, 108	*Edinburgh Castle* (1844) 122	*Jackal* (1875) 78, 128
Aurora (1839) 16, 61, 64, 118	*Elk* (1853) 69-71, 125	*James Watt* (1824) 6, 13, 119
Azalea (1878) 31-33, 38-41, 44, 46, 48, 51, 114	*Elm* (1884) 36, 37, 41, 44, 48, 51, 115	
	Emerald (1855) 17, 112	*Lamprey* (1881) 80, 89, 129
	Enterprize (1826) 59, 119	*Lapwing* (1848) 67, 123
Beagle (1864) 74, 126	*Erin* (1867) 24, 112	*Laurel* (1850) 16-18, 21, 109
Bear (1870) 77, 78, 81, 82, 87, 127	*Ermine* (1912) 98-100, 102, 105, 132	*Laurel* (1850) 124
Beaver (1854) 70, 125		*Laurel* (1863) 22, 109
Bee (1857) 109	*Falcon* (1860) 17, 21, 23, 24, 112	*Laurel* (1864) 24, 27, 29, 112
Belfast (1829) 59, 60, 117	*Fern* (1871) 28, 29, 32, 37, 42, 44, 113	*Laurel* (1873) 40, 115
Bison (1871) 77, 128	*Fern* (1900) 44-46, 48, 53, 54, 116	*Leopard* (1858) 70, 71, 73, 125
Boar (1898) 131	*Ferret* (1871) 77, 128	*Lily* (1896) 42-44, 48, 54, 116
Brier (1882) 33, 38, 44, 48, 114	*Fingal* (1826) 8, 58, 60, 117	*Limpet* (1882) 81, 82, 130
Britannia (1815) 7, 10, 11, 108	*Fire King* (1838) 61, 64, 118	*Liverpool* (1830) 5, 59, 60, 118
Broom (1904) 54, 116	*Fox* (1863) 73, 74, 126	*Lizard* (1881) 80, 129
Buffalo (1865) 75-77, 81, 127	*Fox* (1864) 74, 126	*Llama* (1865) 75, 76, 79, 127
Buzzard (1884) 81, 83, 130	*Foyle* (1829) 11, 12, 108	*Lochfine* (1847) 66, 124
		Locust (1881) 80, 129
Cairnsmore (1896) 55, 116	*Gardenia* (1885) 37, 46, 48, 115	*Londonderry* (1826) 10, 11, 108
Camel (1866) 74-76, 81, 127	*Garland* (1855) 19, 21-23, 25, 109	*Londonderry* (1841) 12-16, 20, 109
Camilla (1849) 65, 67, 69, 124	*Garland* (1863) 27, 28, 113	*Lurcher* (1906) 95, 96, 103, 131
Cardiff Castle (1844) 67, 121	*Gazelle* (1832) 60, 119	*Lurcher* (1922) 102, 132
Cedar (1878) 31-33, 44, 48, 51, 114	*Giraffe* (1860) 72, 73, 126	*Lynx* (1854) 69, 71, 73-75, 125
Clyde (1831) 59, 61, 119	*Glasgow* (1829) 5, 59, 118	*Lyra* (1848) 65, 69, 124
Clydesdale (1826) 11, 14, 108	*Glenalbyn* (1834) 120	
Cobra (1889) 83-85, 130	*Gorilla* (1881) 80, 95, 98, 129	*Magpie* (1898) 89, 90, 99, 103, 131
Coney (1918) 100, 101, 103, 132	*Gorilla* (1922) 102, 103, 132	*Maid of Islay* (1824) 121
Craignish Castle (1844) 67, 122	*Grampus* (1877) 81, 82, 87, 93, 129	*Maid of Morven* (1826) 66, 124
Culloden (1844) 67, 120	*Grouse* (1891) 86, 89, 103, 130	*Maid of Perth* (1847) 67, 123
Culzean (1897) 55, 116		*Manchester* (1832) 59, 61, 119
Curlew (1837) 66, 67, 123	*Hare* (1886) 82, 130	*Maple* (1914) 51-55, 116
Cygnet (1848) 67, 123	*Harrier* (1858) 70, 71, 126	*Mastiff* (1878) 78-82, 85, 87, 93, 128
	Hazel (1907) 48-51, 54, 106, 116	*Mercury* (1835) 60, 118
Daisy (1884) 36, 48, 115	*Helen McGregor* (1835) 62, 64, 120	*Merlin* (1844) 66, 67, 124
Daisy (1895) 43, 44, 53, 116	*Henry Bell* (1823) 6, 13, 20, 118	*Merrick* (1878) 51, 57, 117
Daphne (1883) 33, 34, 37, 114	*Heron* (1860) 71, 74, 126	*Moorfowl* (1919) 101, 103, 132
Dolphin (1844) 64, 67, 120	*Holly* (1875) 30-32, 36-38, 113	*Myrtle* (1854) 18, 19, 21, 109

Myrtle (1859)	20, 109
Myrtle (1865)	22, 23, 26, 28, 29, 109
Northman (1847)	18, 111
Olive (1893)	7, 41, 42, 44, 47, 48, 53, 57, 115
Orion (1847)	5, 65-67, 124
Ostrich (1860)	71, 74, 75, 126
Otter (1855)	70, 125
Owl (1873)	77, 78, 81, 82, 85, 87, 93, 128
Panther (1857)	70, 125
Partridge (1906)	95, 96, 99, 100, 103, 131
Penguin (1864)	73-78, 126
Petrel (1845)	65, 67, 123
Pilot (1844)	65, 67, 122
Pioneer (1844)	65, 67, 122
Plover (1848)	123
Plover (1854)	125
Pointer (1896)	87-89, 91, 92, 99, 103, 130
Prince of Wales (1842)	17, 18, 111
Puma (1899)	97-99, 103, 133
Queen of Beauty (1844)	66, 123
Racoon (1868)	76, 81, 127
Rambler (1845)	12, 15, 20, 109
Rapid (1835)	60, 117
Raven (1868)	76, 77, 81, 127
Redbreast (1908)	96, 99, 100, 103, 131
Redbreast (1920)	100, 102, 103, 132
Rob Roy (1834)	62, 64, 120
Roe (1863)	27, 73, 74, 126
Roe (1864)	74, 126
Rook (1878)	79, 128
Rose (1851)	16, 17, 19, 21, 26, 109
Rose (1867)	24, 26-30, 32, 33, 112
Rose (1883)	33-35, 37, 114
Rose (1902)	44, 45, 48, 53, 116
Rothsay Castle (1837)	67, 121
Rover (1836)	12, 15, 17, 18, 20, 109
Rowan (1914)	51, 53, 55, 56
Sable (1911)	98-100, 103, 132
Saint Columb (1834)	11-13, 15, 16, 20, 108
Scotia (1864)	24, 112
Seal (1877)	81, 82, 87, 93, 129
Setter (1906)	96, 101, 102, 131
Setter (1920)	99, 132
Shamrock (1847)	15, 18, 20, 21, 26, 109
Shamrock (1879)	32, 33, 44, 48, 54, 114
Snipe (1870)	75, 76, 127
Spaniel (1895)	87, 89, 92, 99, 130
Sparrow (1892)	97, 98, 133
Stag (1853)	69-71, 73, 74, 125
Stanley (1864)	39, 40, 115
Stork (1851)	67, 69, 124
Sunbeam (1847)	67, 123
Thetis (1845)	64, 65, 67, 69, 124
Thistle (1848)	13, 15, 16, 19, 109
Thistle (1859)	19-22, 109
Thistle (1863)	22, 109
Thistle (1865)	23, 26, 29-32, 111
Thistle (1884)	36, 40, 44, 48, 115
Tiger (1906)	97-99, 103, 105, 133
Turnberry (1889)	50, 51, 55, 116
Unicorn (1836)	60, 61, 119
Vine (1867)	27, 112
Vine (1878)	31, 32, 113
Viper (1906)	94, 95, 99-102, 131
Vulture (1898)	89, 96, 99, 100, 103-105, 131
Walrus (1878)	78, 80, 128
Wasp (1874)	77, 78, 128
Waterloo (1815)	10, 108
Weasel (1866)	27, 75, 76, 127
William Huskisson (1826)	6, 13, 59, 119
Wolf (1863)	73-76, 126
Woodcock (1906)	56, 94-96, 99, 100, 103, 131
Wren (1885)	97, 99, 132
Zebra (1855)	70, 125